# South Carolina Holt Algebra 1 Standardized Test Practice Workbook

## correlated to

## The South Carolina Algebra 1 Standards

**HOLT, RINEHART AND WINSTON**

A Harcourt Classroom Education Company

**Austin** • New York • Orlando • Atlanta • San Francisco • Boston • Dallas • Toronto • London

# To the Teacher:

Answers to the questions in this booklet are provided for you on the South Carolina Algebra 1 One Stop Planner® CD-ROM.

# To the Student

*South Carolina Algebra 1 Test Practice Booklet* contains a set of test questions in both multiple-choice and extended response formats. A sampling of extended response questions are included to provide practice for the types of extended response questions you may find on the South Carolina standardized math test. Multiple choice questions and a set of quantitative comparison questions are found in standardized test format for each lesson in the *Pupil's Edition*. In addition to the standardized test practice for each lesson, there is a one-page chapter test for each chapter.

Copyright © by Holt, Rinehart and Winston

All rights reserved. No part of this publication may be reproduced or transmitted in any form or by any means, electronic or mechanical, including photocopy, recording, or any information storage and retrieval system, without permission in writing from the publisher.

Teachers may photocopy pages in sufficient quantity for classroom use only, not for resale.

Printed in the United States of America

ISBN 0-03-068973-2

123 179 04 03 02

# Table of Contents

| | |
|---|---|
| Explanation of Correlation | SC1 |
| South Carolina Algebra 1 Standards Correlation | SC2 |
| Algebra 1 Constructed Response Practice | SC7 |
| **Chapter 1** From Patterns to Algebra | 1 |
| **Chapter 2** Operations in Algebra | 7 |
| **Chapter 3** Equations | 14 |
| **Chapter 4** Proportional Reasoning and Statistics | 20 |
| **Chapter 5** Linear Functions | 26 |
| **Chapter 6** Inequalities and Absolute Value | 32 |
| **Chapter 7** Systems of Equations and Inequalities | 37 |
| **Chapter 8** Exponents and Exponential Functions | 43 |
| **Chapter 9** Polynomials and Factoring | 50 |
| **Chapter 10** Quadratic Functions | 58 |
| **Chapter 11** Rational Functions | 64 |
| **Chapter 12** Radical Functions and Coordinate Geometry | 70 |
| **Chapter 13** Probability | 78 |
| **Chapter 14** Functions and Transformations | 83 |

## *Explanation of Correlation*

The following is a correlation of the South Carolina Algebra 1 Standards to **Holt, Rinehart and Winston's South Carolina Holt Algebra 1 Standardized Test Practice Workbook** booklet. Page numbers are given on which treatment of the standards may be found in the practice booklet.

In completing the exercises in **Holt, Rinehart and Winston's Algebra 1 South Carolina Holt Algebra 1 Standardized Test Practice Workbook** booklet, students will demonstrate they have built upon the mathematical understandings that are addressed in prekindergarten through the eighth grade. Students will show they understand and meet South Carolina Algebra 1 standards by using symbolic reasoning to represent mathematical situations, to express generalizations, and to study relationships among quantities; by using functions to represent and model problem situations as well as to analyze and interpret relationships; by setting up equations in a wide range of situations and use a variety of methods to solve them; and by using problem solving, representation, reasoning and proof, language and communication, and connections both within and outside mathematics.

# South Carolina
## *Algebra 1 Standards Correlation*

| South Carolina Algebra 1 Standards | Page References |
|---|---|
| **I. UNDERSTANDING FUNCTIONS** <br> **A. Relationships** | |
| 1. Describe independent and dependent quantities in functional relationships. | 2, 5, 7, 11, 30–36, 43–49, 55–57, 61, 67, 74–75, 80, 96–101 |
| 2. Gather and record data or use data sets to determine functional (systematic) relationships between quantities. | 2, 7, 11, 13–14, 16–20, 23, 33, 36, 38–39, 48–49, 55–57, 61, 65, 74, 89 |
| 3. Describe functional relationships for given problem situations and write equations, inequalities, and recursive relations to answer questions arising from the situations. | 2, 7, 11, 13, 16–20, 23, 38–39, 48–49, 55–57, 74, 78, 86–87 |
| 4. Represent relationships among quantities using concrete models, tables, graphs, diagrams, verbal descriptions, equations, and inequalities including representations involving computer algebra systems, spreadsheets, and graphing calculators. | 2, 4–13, 15–23, 27–33, 35–44, 47–49, 53–57, 59–60, 63–65, 67, 69, 72, 74, 78, 83, 86–87, 91–92, 94, 96–97, 99–100 |
| 5. Make judgments about units of measure and scales within a system and between systems. | 2, 7, 16–18, 21–23, 26–27, 29, 32, 38, 48–49, 55–57, 61, 65, 68, 74, 78, 81, 83, 86–87, 89 |
| 6. Interpret and make inferences from explicit and recursive functional relationships. | 2, 5, 7, 14, 16–22, 30–36, 40–43, 46–49, 55–56, 61, 65–78, 80, 82, 85, 96–101 |
| **B. Linear and Quadratic Functions and Data Representations** | |
| 1. Identify and sketch the general forms of linear ($y = x$) and quadratic ($y = x_2$) parent functions. | 30–36, 43–47, 49, 67, 69, 71–73, 75, 100–101 |
| 2. For a variety of situations, identify and determine reasonable domain and range values for given situations. | 48–49, 55–57, 61, 65, 67–68, 72–73, 75, 86–87, 89, 96–101 |
| 3. Interpret situations in terms of given graphs or create situations that fit given graphs. | 27, 30–33, 35 |
| 4. Represent, display, and interpret data using scatterplots, bar graphs, stem-and-leaf plots, and box-and-whiskers diagrams, including representations on graphing calculators and computers. | 6, 27–29, 92 |
| 5. Write a linear equation that fits a data set, check the model for "goodness of fit," and make predictions using the model. | 6, 17–22, 30–36, 47 |

SC1

# South Carolina
## *Algebra 1 Standards Correlation*

| South Carolina Algebra 1 Standards | Page References |
|---|---|
| **C. Generalizations, Algebraic Symbols, and Matrices** | |
| 1. Read, write, and represent very large and very small numbers in a variety of forms including exponential. | 50–57 |
| 2. Use unit analysis to check measurement computations. | None |
| 3. Given situations, determine patterns and represent generalizations algebraically. | 2–3, 7, 11, 16–21, 22–27, 29, 31–32, 37–39, 41, 48–49, 54–57, 61, 65–68, 74, 78, 80–81, 86–87, 90–95 |
| 4. Use symbolic representation, reasoning, and proof to verify statements about numbers. | This standard is tested throughout the booklet. |
| 5. Recognize and justify the relationship between the magnitude of a number and the application of specific operations. | This standard is tested throughout the booklet. |
| 6. Identify and use properties related to operations with matrices (addition, subtraction, and scalar multiplication) to solve applied problems. | 88, 89 |
| **D. Algebraic Expressions in Problem Solving Situations** | |
| 1. Find specific function values and evaluate expressions. | 2–3, 5, 7, 16–22, 31–32, 41, 43, 45, 49, 55–57, 61, 65, 68, 78, 80–83, 89 |
| 2. Simplify polynomial expressions and perform polynomial arithmetic. | 58, 60 |
| 3. Transform and solve equations and inequalities, factoring as necessary in problem situations. | 16–22, 41, 43, 45, 49, 61, 65 |
| 4. Given a problem situation, determine whether to use a rough estimate, an approximation, or an exact answer. Select a suitable method of computing from techniques such as the use of mental mathematics, paper-and-pencil combinations, calculators, and computers. | None |
| 5. Use supporting data to explain why a solution is mathematically reasonable. | None |
| 6. Use the commutative, associative, and distributive properties to simplify algebraic expressions. | 12, 14–15, 20, 60 |

# South Carolina
## Algebra 1 Standards Correlation

| South Carolina Algebra 1 Standards | Page References |
| --- | --- |
| **II. LINEAR FUNCTIONS** <br> **A. Representations** | |
| 1. Determine whether or not given situations can be represented by linear functions. | 4–7, 30–31, 33–36, 43, 47 |
| 2. Based on the constraints of the problem, determine the domain and range values for linear functions. | 4–5, 30, 32–36 |
| 3. Translate among and use algebraic, tabular, graphical, or verbal descriptions of linear functions using computer algebra systems, spreadsheets, and graphing calculators. | None |
| **B. Interpretations** | |
| 1. Develop the concept of slope as rate of change and determine slope from graphs, tables, and algebraic representations. | 31–36, 40, 47, 97, 99–100 |
| 2. Interpret the meaning of slope and intercepts in situations using data, symbolic representations, or graphs. | 30–36, 43, 47, 97, 99–100 |
| 3. With and without using a graphing calculator, investigate, describe, and predict the effects of changes in m and b on the graph of $y = mx + b$. | 35, 99 |
| 4. Graph and write equations of lines given characteristics such as two points, a point and a slope, or a slope and y-intercept. | 4–5, 7, 30–34 |
| 5. Determine the intercepts of linear functions from graphs, tables, and algebraic representations. | 33–34, 43–47, 49 |
| 6. With and without using a graphing calculator, interpret and predict the effects of changing slope and y-intercept in applied situations. | 35, 99 |
| 7. Relate direct variation to linear functions and solve problems involving proportional change. | 32 |

# South Carolina
## *Algebra 1 Standards Correlation*

| South Carolina Algebra 1 Standards | Page References |
|---|---|
| **C. Equations and Inequalities** | |
| 1. Analyze situations involving linear functions and formulate linear equations or inequalities to solve problems. | 7, 31–32, 36–39, 41, 49 |
| 2. Investigate methods for solving linear equations and inequalities using concrete models, graphs, and the properties of equality; select a method and solve the equations and inequalities. | 30, 34, 37–39, 41–47, 49, 65 |
| 3. Use the commutative, associative, distributive, equality, and identity properties to justify the steps in solving equations and inequalities. | 15, 17, 21 |
| 4. Using concrete models for given contexts, interpret and determine the reasonableness of solutions to linear equations and inequalities. | None |
| **D. Systems of Linear Equations** | |
| 1. Analyze situations and formulate systems of linear equations to solve problems. | 43–49 |
| 2. Solve systems of linear equations using concrete models, graphs, tables, and algebraic methods including computer algebra systems, spreadsheets, and graphing calculators. | 43–49 |
| 3. For given contexts, interpret and determine the reasonableness of solutions to systems of linear equations. | 43–44, 46–47, 49 |

# South Carolina
## Algebra 1 Standards Correlation

| South Carolina Algebra 1 Standards | Page References |
|---|---|
| **III. QUADRATIC AND OTHER FUNCTIONS** <br> **A. Quadratic Functions** | |
| 1. Given the constraints of the problem, determine the domain and range values for quadratic functions. | 67–68, 71–73 |
| 2. With and without using a graphing calculator, investigate, describe, and predict the effects of changes in the coefficient a on the graph of $y = ax^2$. | 97–100 |
| 3. With and without using a graphing calculator, investigate, describe, and predict the effects of changes in the constant c on the graph of $y = x^2 + c$. | 97–100 |
| 4. For problem situations, analyze graphs of quadratic functions and draw conclusions. | 72 |
| 5. Solve quadratic equations using concrete models, tables, graphs, and algebraic methods that include factoring and using the quadratic formula as well as computer algebra systems, spreadsheets, and graphing calculators. | 67–73 |
| 6. Relate the solutions of quadratic equations to the roots of their functions. | 70–71, 73 |
| **B. Other Functions** | |
| 1. Use patterns to generate the laws of exponents and apply the laws of exponents in problem-solving situations. | 50–51, 53, 55–57 |
| 2. Analyze data and represent situations involving inverse variation using concrete models, tables, graphs, or algebraic methods as well as computer algebra systems, spreadsheets, and graphing calculators. | 74–75 |
| 3. Analyze data and represent situations involving exponential growth and decay using concrete models, tables, graphs, or algebraic methods as well as computer algebra systems, spreadsheets, and graphing calculators. | 54–57 |

# Standardized Test Practice
## Algebra 1 Constructed Response Practice

### Chapter 1: From Patterns to Algebra

1. The table gives the winning shot put distances, in meters, from 1980 to 1998.

| Year | Meters | Year | Meters |
|------|--------|------|--------|
| 1980 | 43.9 | 1990 | 46.2 |
| 1981 | 44.3 | 1991 | 46.5 |
| 1982 | 44.6 | 1992 | 46.2 |
| 1983 | 44.3 | 1993 | 47.8 |
| 1984 | 44.7 | 1994 | 47.6 |
| 1985 | 43.6 | 1995 | 49.6 |
| 1986 | 45.1 | 1996 | 48.2 |
| 1987 | 44.9 | 1997 | 50.0 |
| 1988 | 46.7 | 1998 | 49.2 |
| 1989 | 45.9 |  |  |

a. Create a scatter plot and draw a line of best fit.

b. Find the equation of the line in standard form.

c. Describe the correlation between the variables and predict the distance that the shot put will be thrown in the years following 1998.

# Standardized Test Practice
## Algebra 1 Constructed Response Practice

2. Joe made the following table to show the population, p, of his hometown from 1910 to 2000.

| Year | Population (in millions) |
|------|--------------------------|
| 1910 | 0.3 |
| 1920 | 0.5 |
| 1930 | 0.7 |
| 1940 | 0.7 |
| 1950 | 0.9 |
| 1960 | 1.1 |
| 1970 | 1.4 |
| 1980 | 1.6 |
| 1990 | 2.1 |
| 2000 | 2.5 |

a. Sketch a scatter plot. Let $t = 0$ represent 1910. Label the plot with the appropriate values for year and population.

b. Using your scatter plot, determine the type of correlation between $p$ and $t$.

c. Predict the population of Joe's hometown in 2010. Explain the strategy you used to make the prediction.

## Chapter 2: Operations in Algebra

1. For what values of $x$ is each statement true?

   a. $x^2 = x$
   b. $x^2 > x$
   c. $x^2 < x$

2. Indicate whether each statement below is true or false and explain why, using mathematical language.

   a. All real numbers are rational.
   b. All integers are rational numbers.
   c. All whole numbers are integers.

# Standardized Test Practice
## Algebra 1 Constructed Response Practice

**Chapter 3: Equations**

1. Larry goes to the store to buy apples.

    a. He finds that the store is selling six apples for two dollars. If Larry wanted to buy only one apple, how much would it cost?

    b. Larry returns to the store and finds that the apples are on sale at a discounted price of six for one dollar. He decides to wait and come back in two weeks in order to get them for free. Explain why Larry is mistaken in his assumption. Tell how much six apples will cost in two weeks and in four weeks if the store continues to discount their price at the same rate.

2. Luke was asked to demonstrate how to solve the equation $3 + 2(x - 7) = 6x + 5$. Determine if each of Luke's steps is correct or incorrect, **based on the preceding step.** If a step is incorrect, describe the error and explain what Luke should have done instead.

    $$3 + 2(x - 7) = 6x + 5$$
    $$5(x - 7) = 6x + 5$$
    $$5x - 35 = 6x + 5$$
    $$-35 = 11x + 5$$
    $$-30 = 11x$$
    $$-30/11 = x$$
    $$-3\,3/11 = x$$

SC8

# Standardized Test Practice
## Algebra 1 Constructed Response Practice

### Chapter 4: Proportional Reasoning and Statistics

1. Officer Dan was patrolling a neighborhood in his patrol car. During the patrol, a child chasing a ball ran out in front of the car. Officer Dan braked and fortunately did not hit the child. After a serious talk with the child about street safety, Officer Dan continued on his patrol. The graph is a record of the patrol car's speed during the drive.

   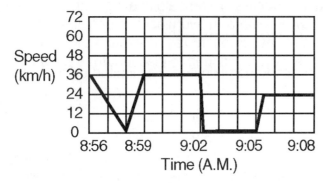

   a. What was the maximum speed of the car during the patrol?

   b. What time was it when Officer Dan hit the brakes to avoid the child?

   c. Describe what could have happened between 8:56 a.m. and 8:59 a.m.

2. On her math exam, Kate received a grade between 48 and 96. Her grade was within 2 standard deviations of the average. What is the range of grades for those students scoring within 1 standard deviation of the average?

SC9

# Standardized Test Practice
## Algebra 1 Constructed Response Practice

### Chapter 6: Inequalities and Absolute Value

1. A farmer figured that the amount of vitamins in a pig's feed should be about 650 mg. The actual amount in each pig's feed may vary up to 30 mg from this amount. Write and solve an absolute-value inequality describing the amount of vitamins in each pig's feed.

### Chapter 7: Systems of Equations and Inequalities

1. A cell phone service offers two rate plans: limited and unlimited. The limited plan is a flat rate of $19.95 per month, plus an additional 20 cents per minute for each minute over 150 for the month. The unlimited plan is a flat rate of $39.95 with unlimited phone time. How much phone time would a person have to use for the limited plan to cost the same as the unlimited plan?

2. The local high school choir is putting on a musical. The choir is selling two types of tickets: adult and student. Adult tickets are $6.25 each, and student tickets are $4.50 each. There are 322 seats in the auditorium. The choir has already spent $525 on props and paint, $62 on advertising in the local paper, and $88 for their choreographer. How many of each type of ticket need to be sold in order for the choir to cover their expenses and make a profit of $900?

# Standardized Test Practice
## *Algebra 1 Constructed Response Practice*

### Chapter 8: Exponents and Exponential Functions

1. Given the expression an, where a is any integer and n is any positive integer, write a rule that can be used to determine the signs of expression. Remember to include all three cases.

2. What are the first four terms of (x + y)7?

### Chapter 12: Radicals, Functions and Coordinate Geometry

Simplify the expression $\dfrac{3}{4 - 3\sqrt{2}} - \dfrac{1}{1 + 4\sqrt{2}}$.

SC11

# Standardized Test Practice
## Algebra 1 Constructed Response Practice

### Chapter 13: Probability

1. Maria plans to go to the movies with her friends but is not sure what she should wear. She has narrowed down her shirt options to a white sweater, a red bouse, and a black long-sleeved shirt. She figures she will wear either a pair of jeans or a pair of khaki pants. As far as shoes, she needs to decide among her new sandals, her black boots, and her favorite pair of tennis shoes.

   a. Draw a tree diagram to illustrate all of Maria's possible outfits. How many possible outfits does she have?

   b. Determine the probability that she will wear the white sweater and khaki pants.

2. There are ten nominees for the high school prom committee.

   a. How many ways are there to choose a committee of 5 people from a group of 10 people?

   b. How many ways are there to choose 3 separate officeholders (chairperson, secretary, and treasurer) from a group of 10 people?

# Standardized Test Practice
## Algebra 1 Constructed Response Practice

**Chapter 14: Functions and Transformations**

A quarterback throws a football from one end of a football field. The ball leaves his hand at a height of 8 feet with an initial velocity of 64 feet per second in the horizontal direction and 48 feet per second in the vertical direction. Given that after $t$ seconds $x(t)$ gives the horizontal distance and $y(t)$ gives the vertical distance, the following parametric equations describe the ball's path.

$$x(t) = 64t$$
$$y(t) = 8 + 48t - t^2$$

a. What maximum height does the football reach on the throw?

b. What horizontal distance does the ball travel before it hits the ground? How much time does it take for the ball to reach this distance?

# Standardized Test Practice
## 1.1 Using Differences to Identify Patterns

**TEST TAKING STRATEGY** Look at each answer choice before selecting one.

1. **Multiple Choice** What is the next term of the sequence, 52, 48, 44, 40, …?
   - Ⓐ 38
   - Ⓑ 36
   - Ⓒ 34
   - Ⓓ 32

2. **Multiple Choice** The third and fourth terms of a sequence are 9 and 14. If the second differences are a constant 1, which of the following is the sequence?
   - Ⓐ 8, 7, 9, 14, 22, …
   - Ⓑ 2, 5, 9, 14, 20, …
   - Ⓒ 5, 6, 9, 14, 21, …
   - Ⓓ 3, 7, 9, 11, 14, …

3. **Multiple Choice** The sequence 17, 20, 23, 26, … has a constant first difference of:
   - Ⓐ 3
   - Ⓑ −3
   - Ⓒ 6
   - Ⓓ 0

4. **Multiple Choice** Which sequence has a first difference of −22?
   - Ⓐ 17, 20, 23, 26, …
   - Ⓑ 111, 133, 155, 177, …
   - Ⓒ 100, 75, 50, 25, …
   - Ⓓ 66, 44, 22, 0, …

5. **Multiple Choice** A conjecture is:
   - Ⓐ a definition.
   - Ⓑ a conclusion based on facts.
   - Ⓒ something derived from a formula.
   - Ⓓ an educated guess based on observations.

6. **Multiple Choice** Which of the following is *not* a problem solving strategy?
   - Ⓐ Solving a simpler problem
   - Ⓑ Working backwards
   - Ⓒ Finding the sum
   - Ⓓ Looking for a pattern

*Quantitative Comparison* In Exercises 7−9, choose the letter of the statement below that is true about the quantities in Columns I and II.

A The number in Column I is greater.
B The number in Column II is greater.
C The two numbers are equal.
D The relationship cannot be determined from the given information.

| | Column I | Column II |
|---|---|---|
| 7. | the first differences in the sequence 22, 25, 28, 31, … | the second differences in the sequence 1, 4, 9, 16, 25, … |
| | Ⓐ   Ⓑ | Ⓒ   Ⓓ |
| 8. | the next term in the sequence | |
| | 1, 12, 29, … | 2, 8, 14, … |
| | Ⓐ   Ⓑ | Ⓒ   Ⓓ |
| 9. | the sum of the next two terms in the sequence | |
| | 10, 17, 26, 37, 50, … | 4, 14, 24, 34, 44, … |
| | Ⓐ   Ⓑ | Ⓒ   Ⓓ |

10. **Multiple Choice** Determine which of the following sequences does *not* have constant first differences.
    - Ⓐ 5, 8, 11, 14, 17, …
    - Ⓑ 87, 72, 57, 42, 27, …
    - Ⓒ 9, 12, 16, 21, 27, 34, …
    - Ⓓ 2, 4, 6, 8, 10, 12, …

Algebra 1

NAME _____ CLASS _____ DATE _____

# Standardized Test Practice
## 1.2 Variables, Expressions, and Equations

**TEST TAKING STRATEGY** Be aware that common mistakes are usually included in the choices.

1. **Multiple Choice** Find the value of the expression $3c + 5$ when $c = 3$.
   - Ⓐ 11
   - Ⓑ 38
   - Ⓒ 24
   - Ⓓ 14

2. **Multiple Choice** Use guess-and-check to solve the equation $11x - 23 = 76$ for $x$.
   - Ⓐ $x = 4$
   - Ⓑ $x = \frac{53}{11}$
   - Ⓒ $x = 5$
   - Ⓓ $x = 9$

3. **Multiple Choice** If CDs cost $15 each and video games cost $30 each, find an expression for the cost of $c$ CDs and $v$ video games.
   - Ⓐ $45cv$
   - Ⓑ $15c + 30v$
   - Ⓒ $15v + 30c$
   - Ⓓ $450cv$

4. **Multiple Choice** Tickets for a baseball game cost $6.50 each. You have $175.50. What equation can be used to determine the number of tickets that can be purchased?
   - Ⓐ $x + 6.5 = 175.5$
   - Ⓑ $x - 6.5 = 175.5$
   - Ⓒ $6.5x = 175.5$
   - Ⓓ $175.5x = 6.5$

5. **Multiple Choice** Find the value of the expression $2w - 7$ when $w = 9$.
   - Ⓐ 25
   - Ⓑ 11
   - Ⓒ 6
   - Ⓓ 22

6. **Multiple Choice** If $n = 4$, what expressions equals 28?
   - Ⓐ $n + 20$
   - Ⓑ $7n$
   - Ⓒ $\frac{21}{n}$
   - Ⓓ $32 - 2n$

*Quantitative Comparison* In Exercises 7–9, choose the letter of the statement below that is true about the quantities in Columns I and II.

A The number in Column I is greater.
B The number in Column II is greater.
C The two numbers are equal.
D The relationship cannot be determined from the given information.

| | Column I | | Column II | |
|---|---|---|---|---|
| 7. | $4x$, when $x = 4$ | | $2x + 3$, when $x = 4$ | |
| | Ⓐ | Ⓑ | Ⓒ | Ⓓ |
| 8. | $3x$, when $x = 12$ | | $2x + 4$, when $x = 15$ | |
| | Ⓐ | Ⓑ | Ⓒ | Ⓓ |
| 9. | the coefficient of the variable in the expression $2x - 3$ | | the constant in the expression $3x + 7$ | |
| | Ⓐ | Ⓑ | Ⓒ | Ⓓ |

10. **Multiple Choice** Pumpkins cost $8 each. Which equation describes the number of pumpkins that you can buy if you have $56?
    - Ⓐ $y = 7$
    - Ⓑ $y = 3$
    - Ⓒ $y = 12$
    - Ⓓ $y = 14$

11. **Multiple Choice** For which value of $w$ is the value of $2w - 4$ equal to 24?
    - Ⓐ $w = 5$
    - Ⓑ $w = 3$
    - Ⓒ $w = 12$
    - Ⓓ $w = 14$

12. **Multiple Choice** Sue charges a flat rate of $5 to plant flowers, and an additional $2 per hour, $h$, for each hour she works. This is best represented by:
    - Ⓐ $(5)(2h)$
    - Ⓑ $5 - 2h$
    - Ⓒ $2h + 5$
    - Ⓓ $5h + 2$

NAME _____ CLASS _____ DATE _____

# Standardized Test Practice
## 1.3 The Algebraic Order of Operations

**TEST TAKING STRATEGY** Use number sense to eliminate unreasonable choices.

1. **Multiple Choice** When finding the value of an expression, which of the following should be performed first?
   - Ⓐ all operations with exponents
   - Ⓑ all additions
   - Ⓒ all multiplications
   - Ⓓ the operation within the innermost grouping symbol

2. **Multiple Choice** For which value of $m$ is the value of $2m + 3$ equal to 19?
   - Ⓐ $m = 14$
   - Ⓑ $m = 8$
   - Ⓒ $m = 7$
   - Ⓓ $m = 4$

3. **Multiple Choice** If $x = 6$, which of the following is the value of $x \cdot (4 + 3)$?
   - Ⓐ 13
   - Ⓑ 27
   - Ⓒ 42
   - Ⓓ 72

4. **Multiple Choice** Which value is equivalent to $7 + 5^2$?
   - Ⓐ 17
   - Ⓑ 24
   - Ⓒ 32
   - Ⓓ 144

5. **Multiple Choice** What is the first step in evaluating $2\{6[2(6 - 4)]\} \div 4$?
   - Ⓐ $6 - 4$
   - Ⓑ $2 \div 4$
   - Ⓒ $2 \times 6$
   - Ⓓ $6(6)$

6. **Multiple Choice** Which expression is equivalent to 0?
   - Ⓐ $4 \cdot 4 \div 4 + 0$
   - Ⓑ $0 \cdot 6 + 1 + (6 \cdot 0)$
   - Ⓒ $1 \div 1 \cdot 0(8 - 4)$
   - Ⓓ $9 - 0 \div 8 + 9$

7. **Multiple Choice** Evaluate $5a + b$ when $a = 2$ and $b = 4$.
   - Ⓐ 14
   - Ⓑ 22
   - Ⓒ 30
   - Ⓓ 40

**Quantitative Comparison** In Exercises 8–11, choose the letter of the statement below that is true about the quantities in Columns I and II.
- **A** The number in Column I is greater.
- **B** The number in Column II is greater.
- **C** The two numbers are equal.
- **D** The relationship cannot be determined from the given information.

| | Column I | | Column II | |
|---|---|---|---|---|
| 8. | $8 + 6 \cdot 5$ | | $(8 + 6) \cdot 5$ | |
| | Ⓐ | Ⓑ | Ⓒ | Ⓓ |
| 9. | $32 \div 4 + 4 \div 2$ | | $(32 \div 4) + (4 \div 2)$ | |
| | Ⓐ | Ⓑ | Ⓒ | Ⓓ |
| 10. | $(10 - 3)^2$ | | $10 - 3^2$ | |
| | Ⓐ | Ⓑ | Ⓒ | Ⓓ |
| 11. | $2x + 4$ | | $3y + 4$ | |
| | Ⓐ | Ⓑ | Ⓒ | Ⓓ |

12. **Multiple Choice** The perimeter of a rectangle is $2(l + w)$, where $l$ is its length and $w$ is its width. What is the perimeter of a rectangle with length 8 and width 3?
    - Ⓐ 11
    - Ⓑ 19
    - Ⓒ 22
    - Ⓓ 48

13. **Multiple Choice** Which of the following statements is true?
    - Ⓐ $48 \div 8 - 4 \div 2 = 24$
    - Ⓑ $48 \div (8 - 4) \div 2 = 24$
    - Ⓒ $48 \div (8 - 4 \div 2) = 24$
    - Ⓓ $48 \div [(8 - 4) \div 2] = 24$

14. **Multiple Choice** Evaluate $56 - 38 \div 2$.
    - Ⓐ 9
    - Ⓑ 14
    - Ⓒ 37
    - Ⓓ 47

NAME _____ CLASS _____ DATE _____

# Standardized Test Practice
## 1.4 Graphing with Coordinates

**TEST TAKING STRATEGY**  Draw a graph or diagram to help you visualize the problem.

1. **Multiple Choice**  In which quadrant of the coordinate plane would you plot the ordered pair $(-7, 3)$?
   - (A) I
   - (B) II
   - (C) III
   - (D) IV

2. **Multiple Choice**  Which point is *not* located in Quadrant IV on the coordinate plane.
   - (A) $(2, -6)$
   - (B) $(5, -5)$
   - (C) $(10, -1)$
   - (D) $(0, 0)$

3. **Multiple Choice**  Determine which set of ordered pairs lie on a straight line.
   - (A) $(-3, 2), (0, 4), (2, 6)$
   - (B) $(-3, -9), (0, -3), (2, 1)$
   - (C) $(0, 7), (0, 0), (3, 4)$
   - (D) $(-5, -3), (2, -1), (-2, -7)$

4. **Multiple Choice**  In the equation $y = \frac{1}{2}x + 3$ you can find the values of $y$ by substituting 2, 4, and 6 for $x$. Which table corresponds to this process?

   (A)
   | x | 2 | 4 | 6 |
   |---|---|---|---|
   | y | 5.5 | 7.5 | 9.5 |

   (B)
   | x | 2 | 4 | 6 |
   |---|---|---|---|
   | y | 5.5 | 7 | 9 |

   (C)
   | x | 2 | 4 | 6 |
   |---|---|---|---|
   | y | 4 | 5 | 6 |

   (D)
   | x | 2 | 4 | 6 |
   |---|---|---|---|
   | y | 4 | 8 | 12 |

**Quantitative Comparison**  In Exercises 5–8, choose the letter of the statement below that is true about the quantities in Columns I and II.

A  The number in Column I is greater.
B  The number in Column II is greater.
C  The two numbers are equal.
D  The relationship cannot be determined from the given information.

**Use the graph for Questions 5–6.**

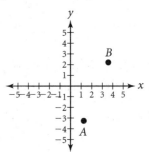

| Column I | Column II |
|---|---|
| 5. the x-coordinate of point A  (A) (B) | the y-coordinate of point A  (C) (D) |
| 6. the x-coordinate of point B  (A) (B) | the y-coordinate of point B  (C) (D) |
| 7. the sum of the coordinates of a point in Quadrant I  (A) (B) | the sum of the coordinates of a point in Quadrant III  (C) (D) |
| 8. the sum of the coordinates of points $(3, -7)$  (A) (B) | the sum of the coordinates of point $(-3, 5)$  (C) (D) |

NAME _____ CLASS _____ DATE _____

# Standardized Test Practice
## 1.5 Representing Linear Patterns

**TEST TAKING STRATEGY**  Check your answers once you complete the test.

1. **Multiple Choice**  Which equation represents the data pattern in the table?

   | 0 | 1 | 2 | 3 | 4 |
   |---|---|---|---|---|
   | −3 | −1 | 1 | 3 | 5 |

   Ⓐ $y = x + 2$
   Ⓑ $y = 2x$
   Ⓒ $y = 2x - 3$
   Ⓓ $y = x - 3$

2. **Multiple Choice**  Which is an ordered pair of the data?

   | 0 | 1 | 2 | 3 | 4 |
   |---|---|---|---|---|
   | 0 | 6 | 12 | 18 | 24 |

   Ⓐ (0, 1)
   Ⓑ (12, 18)
   Ⓒ (1, 6)
   Ⓓ (2, 18)

3. **Multiple Choice**  In the equation $y = 3x + 4$, the "$x$" represents the:

   Ⓐ independent variable.
   Ⓑ dependent variable.
   Ⓒ constant.
   Ⓓ linear equation.

4. **Multiple Choice**  Which equation best represents the data pattern in the table?

   | $x$ | 0 | 1 | 2 | 3 | 4 |
   |---|---|---|---|---|---|
   | $y$ | 250 | 220 | 190 | 160 | 130 |

   Ⓐ $y = -30x + 250$
   Ⓑ $x = -30y + 250$
   Ⓒ $y = 30x - 250$
   Ⓓ $x = 30y - 250$

**Quantitative Comparison**  In Exercises 5–7, choose the letter of the statement below that is true about the quantities in Columns I and II.

**A** The number in Column I is greater.
**B** The number in Column II is greater.
**C** The two numbers are equal.
**D** The relationship cannot be determined from the given information.

| Column I | Column II |
|---|---|

5. the independent variable in the equation $y = 4x - 5$ | the dependent variable in the equation $y = 4x - 5$
   Ⓐ  Ⓑ  Ⓒ  Ⓓ

6. the coefficient of the independent variable in the equation $y = x + 2$ | the coefficient of the dependent variable in the equation $y = x + 2$
   Ⓐ  Ⓑ  Ⓒ  Ⓓ

7. the time it takes to travel 200 miles at $r$ miles per hour | the time it takes to travel 300 miles at $r$ miles per hour
   Ⓐ  Ⓑ  Ⓒ  Ⓓ

8. **Multiple Choice**  Distance traveled is usually represented by the formula:

   Ⓐ $d = \dfrac{r}{t}$
   Ⓑ $d = \dfrac{t}{r}$
   Ⓒ $d = r + t$
   Ⓓ $d = rt$

9. **Multiple Choice**  The graph of $y = 4x + 5$ can best be described as

   Ⓐ a horizontal line
   Ⓑ independent
   Ⓒ a curved line
   Ⓓ a straight line

Algebra 1

# Standardized Test Practice
## 1.6 Scatterplots and Lines of Best Fit

**TEST TAKING STRATEGY** Avoid changing your answer unless you are certain another is better.

1. **Multiple Choice** Which of the following graphs represent a negative correlation.

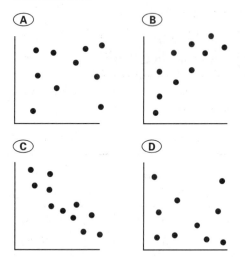

Use the scatter plot below to answer Exercises 2–3.

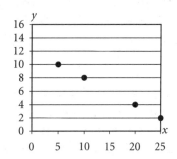

2. **Multiple Choice** Which ordered pair is not represented in the scatter plot?

   Ⓐ (25, 2)   Ⓑ (20, 4)
   Ⓒ (5, 10)   Ⓓ (0, 14)

3. **Multiple Choice** The scatter plot represents what type of correlation?

   Ⓐ strong positive
   Ⓑ strong negative
   Ⓒ little to none
   Ⓓ none of the above

4. **Multiple Choice** Select the graph with the most accurate line of best fit.

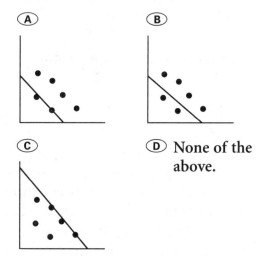

   Ⓓ None of the above.

*Quantitative Comparison* In Exercise 5 choose the letter of the statement below that is true about the quantities in Columns I and II.

**A** The number in Column I is greater.
**B** The number in Column II is greater.
**C** The two numbers are equal.
**D** The relationship cannot be determined from the given information.

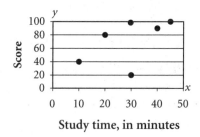

Study time, in minutes

| | Column I | Column II |
|---|---|---|
| 5. | the score that correlates with 20 minutes of studying | the number of minutes studying that correlates with a score of 80 |
| | Ⓐ    Ⓑ | Ⓒ    Ⓓ |

6  Standardized Test Practice 1.6          Algebra 1

NAME _____ CLASS _____ DATE _____

# SAT/ACT Chapter Test

## Chapter 1  From Patterns to Algebra

**TEST TAKING STRATEGY**  Know what each question is asking.

1. **Multiple Choice**  In which quadrant is the ordered pair $(-4, -2)$ located?
   - Ⓐ I
   - Ⓑ II
   - Ⓒ III
   - Ⓓ IV

2. **Multiple Choice**  If $m = 4$, $5m - 3$ equals:
   - Ⓐ 51
   - Ⓑ 23
   - Ⓒ 17
   - Ⓓ 54

3. **Multiple Choice**  The cost, $c$, of renting a video game machine is $9.95 plus $4 per game, $g$. Which equation represents the cost of renting a video game machine and $g$ games?
   - Ⓐ $c = 4g(9.95)$
   - Ⓑ $c = 9.9g + 4$
   - Ⓒ $c = 13.95g$
   - Ⓓ $c = 4g + 9.95$

4. **Multiple Choice**  What are the next two terms in the sequence 5, 7, 11, 17, 25, … ?
   - Ⓐ 33, 43
   - Ⓑ 35, 45
   - Ⓒ 33, 45
   - Ⓓ 35, 47

5. **Multiple Choice**  A scatter plot showing the relationship of shoe size to time spent on the telephone has what type of correlation?
   - Ⓐ positive
   - Ⓑ strong negative
   - Ⓒ weak negative
   - Ⓓ little to none

6. **Multiple Choice**  When evaluating the expression $4 \cdot 3^2 - 24 \div 4 \cdot 3$, which operation should be performed first?
   - Ⓐ $4 \cdot 3$
   - Ⓑ $-24 \div 4$
   - Ⓒ $3^2$
   - Ⓓ $3^2 - 24$

7. **Multiple Choice**  Which ordered pair does *not* lie on the same line as the others?
   - Ⓐ $(-3, -2)$
   - Ⓑ $(1, 0)$
   - Ⓒ $(5, 2)$
   - Ⓓ $(3, 2)$

**Quantitative Comparison**  In Exercises 8–10, choose the letter of the statement below that is true about the quantities in Columns I and II.

A  The number in Column I is greater.
B  The number in Column II is greater.
C  The two numbers are equal.
D  The relationship cannot be determined from the given information.

| | Column I | Column II |
|---|---|---|
| 8. | $24 - 12 \div 12$ | $3x - 4$ when $x = 7$ |
| | Ⓐ  Ⓑ | Ⓒ  Ⓓ |
| 9. | the $x$-coordinate of (3, 5) | the $y$-coordinate of (9, 4) |
| | Ⓐ  Ⓑ | Ⓒ  Ⓓ |
| 10. | The second differences in the sequence 4, 6, 9, 13, 18, … | The first differences in the sequence 12, 16, 20, 24, 28, … |
| | Ⓐ  Ⓑ | Ⓒ  Ⓓ |

11. **Multiple Choice**  A local video store charges a $9.50 membership fee in addition to the cost of each video rented, $n$. Which equation represents the data in the table?

| Number of videos rented, $n$ | 0 | 1 | 2 | 3 | 4 |
|---|---|---|---|---|---|
| Cost, $c$, in dollars | 9.5 | 13 | 16.5 | 20 | 23.5 |

   - Ⓐ $c = 9.5n + 3$
   - Ⓑ $c = 3.5n + 9.5$
   - Ⓒ $n = 9.5c + 3$
   - Ⓓ $n = 3.5c + 9.5$

12. **Multiple Choice**  What is the last step in the algebraic order of operations?
    - Ⓐ exponents
    - Ⓑ addition
    - Ⓒ grouping symbols
    - Ⓓ division

Algebra 1                                   SAT/ACT Chapter Test Chapter 1      7

NAME _____ CLASS _____ DATE _____

# Standardized Test Practice
## 2.1 The Real Numbers and Absolute Value

**TEST TAKING STRATEGY** Use number sense to eliminate unreasonable choices.

1. **Multiple Choice** Which number set best describes the set to which $\sqrt{5}$ belongs?
   - Ⓐ real numbers
   - Ⓑ irrational numbers
   - Ⓒ integers
   - Ⓓ whole numbers

2. **Multiple Choice** Which symbol makes $-0.4 \_\_\_ -0.15$ a true statement?
   - Ⓐ $<$
   - Ⓑ $>$
   - Ⓒ $=$
   - Ⓓ none of the above

3. **Multiple Choice** Which statement is false?
   - Ⓐ $0 < |-10|$
   - Ⓑ $-16 < -32$
   - Ⓒ $\frac{2}{3} = 0.\overline{66}$
   - Ⓓ $-0 = 0$

4. **Multiple Choice** Which expression is equal to $-14$?
   - Ⓐ $|14| + 0$
   - Ⓑ $-|-30| + 16$
   - Ⓒ $|8 - 22|$
   - Ⓓ $|-7| - |7|$

5. **Multiple Choice** Simplify $|-6| + |16|$.
   - Ⓐ 10
   - Ⓑ $-10$
   - Ⓒ 22
   - Ⓓ $-22$

6. **Multiple Choice** What is the opposite of $-36$, minus the absolute value of $-36$?
   - Ⓐ 0
   - Ⓑ 36
   - Ⓒ $-36$
   - Ⓓ $-72$

7. **Multiple Choice** What expression shows the absolute value of 12.5 plus the absolute value of $-4$?
   - Ⓐ $|12 + 5| + |-4|$
   - Ⓑ $|-12.5| - |4|$
   - Ⓒ $|-12.5| + 4$
   - Ⓓ $-12.5 + 4$

**Quantitative Comparison** In Exercises 8–11, choose the letter of the statement below that is true about the quantities in Columns I and II.

A The number in Column I is greater.
B The number in Column II is greater.
C The two numbers are equal.
D The relationship cannot be determined from the given information.

| | Column I | Column II | |
|---|---|---|---|
| 8. | the absolute value of 17 | the opposite of 17 | |
| | Ⓐ Ⓑ | Ⓒ Ⓓ | |
| 9. | the opposite of 4 | the opposite of $-\frac{3}{4}$ | |
| | Ⓐ Ⓑ | Ⓒ Ⓓ | |
| 10. | $|8 - 3|$ | $|4.9|$ | |
| | Ⓐ Ⓑ | Ⓒ Ⓓ | |
| 11. | $-(-10)$ | $-|-10|$ | |
| | Ⓐ Ⓑ | Ⓒ Ⓓ | |

12. **Multiple Choice** Which of the following is *not* a repeating decimal?
   - Ⓐ $0.\overline{26}$
   - Ⓑ $0.81813813$
   - Ⓒ $\frac{1}{9}$
   - Ⓓ $-0.16341634\ldots$

13. **Multiple Choice** To which number set(s) does $-10$ belong?
   - Ⓐ Irrational Numbers
   - Ⓑ Integers
   - Ⓒ Whole Numbers
   - Ⓓ Natural Numbers

NAME _____ CLASS _____ DATE _____

# Standardized Test Practice
## 2.2 Adding Real Numbers

**TEST TAKING STRATEGY**  When you find the answer, remember to check your work.

1. *Multiple Choice*  What is the sum of $-8 + (-22)$?
   - Ⓐ 14
   - Ⓑ $-14$
   - Ⓒ 30
   - Ⓓ $-30$

2. *Multiple Choice*  The value 32 is the result of which number sentence?
   - Ⓐ $62 - 20$
   - Ⓑ $-94 + (-62)$
   - Ⓒ $47 + (-15)$
   - Ⓓ $-32 - 32$

3. *Multiple Choice*  The expression, $0.28 + (-1.4)$, equals:
   - Ⓐ $-0.42$
   - Ⓑ $-1.68$
   - Ⓒ $-1.32$
   - Ⓓ $-1.12$

4. *Multiple Choice*  What is the sum of $\frac{-1}{2} + \frac{1}{3}$?
   - Ⓐ 0
   - Ⓑ $-10$
   - Ⓒ $-\frac{1}{6}$
   - Ⓓ $-\frac{1}{5}$

5. *Multiple Choice*  Evaluate $a + (-c)$, when $a = -2$ and $c = -3$.
   - Ⓐ $-1$
   - Ⓑ $-5$
   - Ⓒ 1
   - Ⓓ 5

6. *Multiple Choice*  If $a = 1$, $b = -3$, and $c = -4$, which value is equivalent to $(b - a) + (-c)$?
   - Ⓐ 0
   - Ⓑ $-6$
   - Ⓒ 2
   - Ⓓ $-8$

7. *Multiple Choice*  Which number sentence is true when $a = 3$, $b = 2$, and $c = -9$, is substituted into the expression $2c + 4ab$?
   - Ⓐ $18 - 24$
   - Ⓑ $4 + (-92)$
   - Ⓒ $6 + 72$
   - Ⓓ $-18 + 24$

*Quantitative Comparison*  In Exercises 8–11, choose the letter of the statement below that is true about the quantities in Columns I and II.

**A**  The number in Column I is greater.
**B**  The number in Column II is greater.
**C**  The two numbers are equal.
**D**  The relationship cannot be determined from the given information.

| | Column I | Column II |
|---|---|---|
| 8. | $-5 + 5$ | $28 + (-17)$ |
|    | Ⓐ Ⓑ | Ⓒ Ⓓ |
| 9. | $-13 + 20$ | $13 - 20$ |
|    | Ⓐ Ⓑ | Ⓒ Ⓓ |
| 10. | $-\frac{3}{5} + \frac{1}{4}$ | $\frac{2}{3} + \frac{1}{5}$ |
|    | Ⓐ Ⓑ | Ⓒ Ⓓ |
| 11. | $7 + (-7)$ | $3.2 + (-5.1) + 1.9$ |
|    | Ⓐ Ⓑ | Ⓒ Ⓓ |

12. *Multiple Choice*  Which property is used in $-3.6 + 0 = -3.6$?
   - Ⓐ Identity Property for Addition
   - Ⓑ Rules for Adding Signed Numbers
   - Ⓒ Additive Inverse Property
   - Ⓓ Definition of Absolute Value

13. *Multiple Choice*  $-23 + 23 = 0$ is an example of which property?
   - Ⓐ Identity Property for Addition
   - Ⓑ Rules for Adding Signed Numbers
   - Ⓒ Additive Inverse Property
   - Ⓓ Definition of Absolute Value

Algebra 1

Standardized Test Practice 2.2

# Standardized Test Practice
## 2.3 Subtracting Real Numbers

**TEST TAKING STRATEGY** Turn a multiple choice question into a true/false question.

1. **Multiple Choice** What value is the difference between 8 and $-22$?
   - Ⓐ 14
   - Ⓑ $-14$
   - Ⓒ 30
   - Ⓓ $-30$

2. **Multiple Choice** Evaluate $x - (-y)$ when $x = -4$, and $y = -3$.
   - Ⓐ 1
   - Ⓑ $-1$
   - Ⓒ 7
   - Ⓓ $-7$

3. **Multiple Choice** What is $-3 - 7$?
   - Ⓐ 10
   - Ⓑ $-10$
   - Ⓒ 4
   - Ⓓ $-4$

4. **Multiple Choice** Which is *not* equivalent to $-4 - (-11)$?
   - Ⓐ $-4 - 11$
   - Ⓑ 7
   - Ⓒ $-4 + 11$
   - Ⓓ $11 - 4$

5. **Multiple Choice** Which number line has a distance of 5 between points $A$ and $B$?

6. **Multiple Choice** If $x = 4$ and $y = -2$ what does $(x - y)$ equal?
   - Ⓐ 2
   - Ⓑ $-2$
   - Ⓒ 6
   - Ⓓ $-6$

**Quantitative Comparison** In Exercises 7–11, choose the letter of the statement below that is true about the quantities in Columns I and II.

**A** The number in Column I is greater.
**B** The number in Column II is greater.
**C** The two numbers are equal.
**D** The relationship cannot be determined from the given information.

| | Column I | | | Column II | |
|---|---|---|---|---|---|
| 7. | $7 - 10$ | | | $-10 - 7$ | |
| | Ⓐ | Ⓑ | Ⓒ | Ⓓ | |
| 8. | $8 - (-4)$ | | | $6 + 6$ | |
| | Ⓐ | Ⓑ | Ⓒ | Ⓓ | |
| 9. | $-7 - x$, when $x = 7$ | | | $7 - (-7)$ | |
| | Ⓐ | Ⓑ | Ⓒ | Ⓓ | |
| 10. | $-14 + y$ | | | $y - 16$ | |
| | Ⓐ | Ⓑ | Ⓒ | Ⓓ | |
| 11. | $\dfrac{4}{5} - \dfrac{1}{2}$ | | | $-\dfrac{3}{7} + \dfrac{1}{3}$ | |
| | Ⓐ | Ⓑ | Ⓒ | Ⓓ | |

12. **Multiple Choice** What is the distance between $-23$ and $-49$ on a number line?
    - Ⓐ $d = 26$
    - Ⓑ $d = -26$
    - Ⓒ $d = 72$
    - Ⓓ $d = -72$

13. **Multiple Choice** Which fraction is equivalent to $-\dfrac{2}{3} - \dfrac{1}{4}$?
    - Ⓐ $\dfrac{-3}{7}$
    - Ⓑ $\dfrac{-5}{7}$
    - Ⓒ $\dfrac{-3}{12}$
    - Ⓓ $\dfrac{-11}{12}$

NAME _____ CLASS _____ DATE _____

# Standardized Test Practice
## 2.4 Multiplying and Dividing Real Numbers

**TEST TAKING STRATEGY** Look at all possible answers before choosing one.

1. **Multiple Choice** Simplify $\frac{15}{-5}$.
   - Ⓐ 10
   - Ⓑ 3
   - Ⓒ −10
   - Ⓓ −3

2. **Multiple Choice** What equals $\frac{0}{-8}$?
   - Ⓐ 0
   - Ⓑ 8
   - Ⓒ −8
   - Ⓓ undefined

3. **Multiple Choice** The product of $(-2.7)(-0.3)$ is:
   - Ⓐ less than zero
   - Ⓑ positive
   - Ⓒ a whole number
   - Ⓓ negative

4. **Multiple Choice** What expression is equal to $\frac{-5+7}{4}$?
   - Ⓐ $(-5+7) \div 4$
   - Ⓑ $\frac{-5}{4} - \frac{7}{4}$
   - Ⓒ $4(-5+7)$
   - Ⓓ $-5 + 7 \div 4$

5. **Multiple Choice** Identify the property shown in $7 \cdot \frac{1}{7} = 1$.
   - Ⓐ Identity Property of Multiplication
   - Ⓑ Multiplicative Inverse Property
   - Ⓒ Property of Zero
   - Ⓓ Rules for multiplying signed numbers

6. **Multiple Choice** If $k$ is 5, what does $12(k-3)$ equal?
   - Ⓐ 6
   - Ⓑ 12
   - Ⓒ 24
   - Ⓓ 57

*Quantitative Comparison* In Exercises 7–10, choose the letter of the statement below that is true about the quantities in Columns I and II.

A The number in Column I is greater.
B The number in Column II is greater.
C The two numbers are equal.
D The relationship cannot be determined from the given information.

| | Column I | | Column II | |
|---|---|---|---|---|
| 7. | the product of two negative numbers | | the quotient of two negative numbers | |
| | Ⓐ | Ⓑ | Ⓒ | Ⓓ |
| 8. | $-\frac{12}{6}$ | | $4(-2)$ | |
| | Ⓐ | Ⓑ | Ⓒ | Ⓓ |
| 9. | $8 - 10$ | | $-3 - (-1)$ | |
| | Ⓐ | Ⓑ | Ⓒ | Ⓓ |
| 10. | $\frac{8(-1)}{2-6}$ | | $\frac{14}{x}$ when $x = 7$ | |
| | Ⓐ | Ⓑ | Ⓒ | Ⓓ |

11. **Multiple Choice** Simplify $\left(\frac{7}{5}\right) \div \left(\frac{-5}{7}\right)$.
    - Ⓐ 1
    - Ⓑ −1
    - Ⓒ $\frac{49}{25}$
    - Ⓓ $-\frac{49}{25}$

12. **Multiple Choice** Sam invested $50 in a fund. Every month, for 4 months, he invested another $100. At the end of 4 months he withdrew $\frac{1}{4}$ of the total amount. What amount remained in his account?
    - Ⓐ $450
    - Ⓑ $337.50
    - Ⓒ $400
    - Ⓓ $112.50

Algebra 1  Standardized Test Practice 2.4

# Standardized Test Practice
## 2.5 Properties and Mental Computation

**TEST TAKING STRATEGY** Eliminate choices that are obviously not possible.

1. **Multiple Choice** Identify the property used in $7 + (3 \cdot 2) = 7 + (2 \cdot 3)$.
   - Ⓐ Commutative Property of Addition
   - Ⓑ Commutative Property of Multiplication
   - Ⓒ Associative Property of Addition
   - Ⓓ Associative Property of Multiplication

2. **Multiple Choice** Which is an example of the Associative Property of Multiplication?
   - Ⓐ $8 + 2a = (8 + 2)a$
   - Ⓑ $8 \cdot 2a = (8 + 2)a$
   - Ⓒ $8 \cdot 2a = (2a \cdot 8a)$
   - Ⓓ $8 \cdot 2a = (8 \cdot 2)a$

3. **Multiple Choice** Identify the property used in $4(20 + 8) = 4 \cdot 20 + 4 \cdot 8$.
   - Ⓐ Commutative Property of Addition
   - Ⓑ Commutative Property of Multiplication
   - Ⓒ Distributive Property
   - Ⓓ Associative Property of Multiplication

4. **Multiple Choice** $7[6 + (-6)] = 7(0)$ is an example of which property?
   - Ⓐ Additive Inverse
   - Ⓑ Additive Identity
   - Ⓒ Multiplicative Inverse
   - Ⓓ Multiplicative Identity

5. **Multiple Choice** Identify the property of equality used in "If $5 = 3 + 2$, then $3 + 2 = 5$."
   - Ⓐ Reflexive Property of Equality
   - Ⓑ Symmetric Property of Equality
   - Ⓒ Transitive Property of Equality
   - Ⓓ Substitution Property of Equality

**Quantitative Comparison** In Exercises 6–9, choose the letter of the statement below that is true about the quantities in Columns I and II.

A  The number in Column I is greater.
B  The number in Column II is greater.
C  The two numbers are equal.
D  The relationship cannot be determined from the given information.

| | Column I | Column II | |
|---|---|---|---|
| 6. | the product of two negative numbers | the product of two positive numbers | |
| | Ⓐ    Ⓑ | Ⓒ    Ⓓ | |
| 7. | $-2(4)$ | $4(-2)$ | |
| | Ⓐ    Ⓑ | Ⓒ    Ⓓ | |
| 8. | $-3(0)$ | $-3 - (-1)$ | |
| | Ⓐ    Ⓑ | Ⓒ    Ⓓ | |
| 9. | $12(9 + 3)$ | $12(9) + 12(3)$ | |
| | Ⓐ    Ⓑ | Ⓒ    Ⓓ | |

**Multiple Choice** For Items 10–13, identify the property used in each step of evaluating $(42 + 15) + 58$.
   - Ⓐ Commutative Property of Addition
   - Ⓑ Associative Property of Addition
   - Ⓒ Distributive Property
   - Ⓓ Substitution Property

10. $= (15 + 42) + 58$
    Ⓐ   Ⓑ   Ⓒ   Ⓓ
11. $= 15 + (42 + 58)$
    Ⓐ   Ⓑ   Ⓒ   Ⓓ
12. $= 15 + 100$
    Ⓐ   Ⓑ   Ⓒ   Ⓓ

NAME _____ CLASS _____ DATE _____

# Standardized Test Practice
## 2.6 Adding and Subtracting Expressions

**TEST TAKING STRATEGY** Substitute values for variables to determine if a statement is true.

1. *Multiple Choice* Simplify $14x - 5x$.
   - Ⓐ $11x$
   - Ⓑ $9x$
   - Ⓒ $9x^2$
   - Ⓓ $9$

2. *Multiple Choice* What is the opposite of the expression $8w - (4 - w)$?
   - Ⓐ $-8w + 4 - w$
   - Ⓑ $8w + (4 - w)$
   - Ⓒ $8w - (4 - w)$
   - Ⓓ $-8w + (4 + w)$

3. *Multiple Choice* Which of the following expressions is equivalent to $(-4x + 2y) + (-7x - 3y)$?
   - Ⓐ $-11x - y$
   - Ⓑ $-11x - 5y$
   - Ⓒ $-11x + y$
   - Ⓓ $-11x + 5y$

4. *Multiple Choice* Simplify:
   $(-12x - 5y) - (6x + 7y)$
   - Ⓐ $-6x + 2y$
   - Ⓑ $-18x + 12y$
   - Ⓒ $-6x - 12y$
   - Ⓓ $-18x - 12y$

5. *Multiple Choice* If you simplify $(4.2c - 4d) + (3c - 3.6d)$, what is the coefficient of the $c$ term?
   - Ⓐ $-1.2$
   - Ⓑ $4.5$
   - Ⓒ $1.2$
   - Ⓓ $7.2$

6. *Multiple Choice* Simplify:
   $\left(\dfrac{2x}{3} + 1\right) - \left(\dfrac{x}{4} - 1\right)$
   - Ⓐ $x + 2$
   - Ⓑ $\dfrac{5x}{12} + 2$
   - Ⓒ $\dfrac{11x}{12} + 2$
   - Ⓓ $\dfrac{11x}{12}$

7. *Multiple Choice* In the expression $(2y + 7) - (-2y + 7)$, the constant is:
   - Ⓐ $4$
   - Ⓑ $-4$
   - Ⓒ $14$
   - Ⓓ $0$

*Quantitative Comparison* In Exercises 8–11, choose the letter of the statement below that is true about the quantities in Columns I and II.

A The number in Column I is greater.
B The number in Column II is greater.
C The two numbers are equal.
D The relationship cannot be determined from the given information.

| | Column I | Column II |
|---|---|---|
| 8. | the perimeter of | |

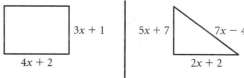

| | Ⓐ | Ⓑ | Ⓒ | Ⓓ |
|---|---|---|---|---|
| 9. | $-y + 2$ | | $-4y + 1$ | |
| | Ⓐ | Ⓑ | Ⓒ | Ⓓ |
| 10. | $5(a + b)$ | | $5a + 5b$ | |
| | Ⓐ | Ⓑ | Ⓒ | Ⓓ |
| 11. | $8d - (4 - d)$ | | $(5 + 6d) + (3d - 5)$ | |
| | Ⓐ | Ⓑ | Ⓒ | Ⓓ |

12. *Multiple Choice* What expression can be used to determine the perimeter of the rectangle?

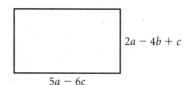

   - Ⓐ $7a - 4b - 5c$
   - Ⓑ $14a - 8b - 10c$
   - Ⓒ $-2abc$
   - Ⓓ $10a^2 - 4b - 6c^2$

Algebra 1      Standardized Test Practice 2.6    **13**

# Standardized Test Practice
## 2.7 Multiplying and Dividing Expressions

**TEST TAKING STRATEGY** Read the directions carefully.

1. **Multiple Choice** If the length of a rectangle is $2x$ and the width is $(4x - 7)$, which expression defines the area?
   - Ⓐ $6x - 14x$
   - Ⓑ $8x^2 - 14x$
   - Ⓒ $6x^2 - 14x$
   - Ⓓ $8x^2 - 7$

2. **Multiple Choice** What property do you use to simplify the expression $-2(8x - 4)$?
   - Ⓐ Commutative
   - Ⓑ Property of Zero
   - Ⓒ Distributive
   - Ⓓ Associative

3. **Multiple Choice** What expression is equivalent to $(4x + 2y - 5) - 4(2y + 1)$?
   - Ⓐ $4x - 6y + 4$
   - Ⓑ $4x - 16y + 4$
   - Ⓒ $4x - 6y - 9$
   - Ⓓ $4x - 16y - 9$

4. **Multiple Choice** What is the quotient of $(7x - 7) \div 7x$?
   - Ⓐ $7 - x$
   - Ⓑ $1 - x$
   - Ⓒ $1 - \dfrac{1}{x}$
   - Ⓓ $49x^2 - 49x$

5. **Multiple Choice** Which expression is *not* equivalent to $\dfrac{8x^2 + 4}{2}$?
   - Ⓐ $4x^2 + 4$
   - Ⓑ $\dfrac{8x^2}{2} + \dfrac{4}{2}$
   - Ⓒ $4x^2 + 2$
   - Ⓓ $\dfrac{1}{2}(8x^2 + 4)$

6. **Multiple Choice** Which expression is $\dfrac{-15x - 5}{-5}$ in simplified form?
   - Ⓐ $-3x + 1$
   - Ⓑ $3x + 1$
   - Ⓒ $-3x$
   - Ⓓ $3x$

7. **Multiple Choice** 7 to the fifth power can be written as:
   - Ⓐ $7 \times 7 \times 7$
   - Ⓑ $7 \times 5$
   - Ⓒ $5^7$
   - Ⓓ $7^5$

**Quantitative Comparison** In Exercises 8–11, choose the letter of the statement below that is true about the quantities in Columns I and II.

**A** The number in Column I is greater.
**B** The number in Column II is greater.
**C** The two numbers are equal.
**D** The relationship cannot be determined from the given information.

| | Column I | | Column II | |
|---|---|---|---|---|
| 8. | $-8x \cdot 3x$ | | $4x \cdot (-6x)$ | |
| | Ⓐ | Ⓑ | Ⓒ | Ⓓ |
| 9. | $-2 \cdot 4x$ | | $7 \cdot 3x$ | |
| | Ⓐ | Ⓑ | Ⓒ | Ⓓ |
| 10. | $\dfrac{4y^2 + 4}{4}$ | | $5y^2 - (4y^2 - 1)$ | |
| | Ⓐ | Ⓑ | Ⓒ | Ⓓ |
| 11. | 5 to the fourth power | | $y$ cubed | |
| | Ⓐ | Ⓑ | Ⓒ | Ⓓ |

12. **Multiple Choice** $-3x^2 + 5x - 2$ is the simplified version of which expression?
   - Ⓐ $\dfrac{15x^2 + 25x + 10}{5x}$
   - Ⓑ $\dfrac{18x^2 - 30x + 12}{6}$
   - Ⓒ $\dfrac{12x^2 - 20x + 8}{-4}$
   - Ⓓ $-6x^2 - 10x - 4$

13. **Multiple Choice** What value is an exponent in the expression $8x - (4 - 3x^2)$?
   - Ⓐ 8
   - Ⓑ 4
   - Ⓒ $-3$
   - Ⓓ 2

# SAT/ACT Chapter Test
## Chapter 2  Operations in Algebra

**TEST TAKING STRATEGY**  Be comfortable and alert for tests.

1. **Multiple Choice** Which is true?
   I. $-2 < 7$   II. $-2 < -7$   III. $2 < -7$
   - Ⓐ only I
   - Ⓑ only I and II
   - Ⓒ only II and III
   - Ⓓ I, II, and III

2. **Multiple Choice** Given $m = -4$, which expression has the same value as $-|m|$?
   - Ⓐ $-m$
   - Ⓑ $-(-m)$
   - Ⓒ $|m|$
   - Ⓓ $|-m|$

3. **Multiple Choice** The value 1.2 is the _____ of $-3.9$ and $5.1$.
   - Ⓐ product
   - Ⓑ quotient
   - Ⓒ difference
   - Ⓓ sum

4. **Multiple Choice** What is the difference of $-11.5$ from 3?
   - Ⓐ $-8.5$
   - Ⓑ $-7.5$
   - Ⓒ $11.8$
   - Ⓓ $14.5$

5. **Multiple Choice** The product of $(-9)$ and $(-7)$ is:
   - Ⓐ $-63$
   - Ⓑ $-16$
   - Ⓒ $16$
   - Ⓓ $63$

6. **Multiple Choice** What is the reciprocal of $-1\frac{3}{4}$?
   - Ⓐ $-2\frac{1}{3}$
   - Ⓑ $-\frac{4}{7}$
   - Ⓒ $-\frac{3}{7}$
   - Ⓓ $1\frac{3}{4}$

7. **Multiple Choice** Find the quotient of $-\frac{1}{6} \div \frac{3}{10}$.
   - Ⓐ $-1\frac{4}{5}$
   - Ⓑ $-\frac{5}{9}$
   - Ⓒ $-\frac{1}{20}$
   - Ⓓ $\frac{1}{20}$

**Quantitative Comparison** In Exercises 8–10, choose the letter of the statement below that is true about the quantities in Columns I and II.

A  The number in Column I is greater.
B  The number in Column II is greater.
C  The two numbers are equal.
D  The relationship cannot be determined from the given information.

| | Column I | Column II |
|---|---|---|
| 8. | the opposite of the negative number, $x$ | the absolute value of the negative number, $x$ |
|   | Ⓐ  Ⓑ | Ⓒ  Ⓓ |
| 9. | the sum of two negative numbers, $x$ and $y$ | the sum of two positive numbers, $x$ and $y$ |
|   | Ⓐ  Ⓑ | Ⓒ  Ⓓ |
| 10. | the product of two negative numbers, $x$ and $y$ | the product of two positive numbers, $x$ and $y$ |
|   | Ⓐ  Ⓑ | Ⓒ  Ⓓ |

11. **Multiple Choice** Which property is illustrated by $3 \cdot (2 \cdot z) = (3 \cdot 2) \cdot z$?
    - Ⓐ Associative Property of Multiplication
    - Ⓑ Commutative Property of Multiplication
    - Ⓒ Distributive Property
    - Ⓒ Transitive Property of Equality

12. **Multiple Choice** Which expression is equivalent to $5c - (c - 3)$?
    - Ⓐ $8$
    - Ⓑ $4c - 3$
    - Ⓒ $4c + 3$
    - Ⓓ $-5c^2 + 15c$

NAME _____ CLASS _____ DATE _____

# Standardized Test Practice
## 3.1 Solving Equations by Adding and Subtracting

**TEST TAKING STRATEGY** Use number sense to eliminate unreasonable choices.

1. **Multiple Choice** Which equation has $-1.9$ as a solution?

   Ⓐ $x + 1.6 = 3.5$ Ⓑ $x - 1.9 = 0$
   Ⓒ $x - 3.5 = 1.6$ Ⓓ $x + 3.5 = 1.6$

2. **Multiple Choice** Carole has $139. If she spent $58 at the shoe store, how much money does she have left?

   Ⓐ $197 Ⓑ $187
   Ⓒ $-81 Ⓓ $81

3. **Multiple Choice** What is the solution to the equation $w + \frac{2}{3} = \frac{4}{9}$?

   Ⓐ $w = \frac{1}{3}$ Ⓑ $w = -\frac{2}{9}$
   Ⓒ $w = \frac{1}{2}$ Ⓓ $w = -\frac{1}{3}$

4. **Multiple Choice** To find the measure of an angle that is supplementary to an angle with a measure of 45°, which is the correct equation?

   Ⓐ $a + 45 = 180$ Ⓑ $a - 45 = 180$
   Ⓒ $a + 45 = 90$ Ⓓ $a - 45 = 90$

5. **Multiple Choice** What is the solution to the equation $8 - k = 15$?

   Ⓐ $k = 23$ Ⓑ $k = -23$
   Ⓒ $k = 7$ Ⓓ $k = -7$

6. **Multiple Choice** What is the first step in solving the equation $x + 57 = 42$?

   Ⓐ Add 42 to both sides.
   Ⓑ Add 57 to both sides.
   Ⓒ Subtract 42 from both sides.
   Ⓓ Subtract 57 from both sides.

*Quantitative Comparison* In Exercises 7–9, choose the letter of the statement below that is true about the quantities in Columns I and II.

A The number in Column I is greater.
B The number in Column II is greater.
C The two numbers are equal.
D The relationship cannot be determined from the given information.

| | Column I | Column II |
|---|---|---|
| 7. | the solution to $c - 14 = -9$ | the solution to $c + 5 = -4$ |
| | Ⓐ    Ⓑ | Ⓒ    Ⓓ |
| 8. | the solution to $-4 - f = 7$ | the solution to $15 - f = 8$ |
| | Ⓐ    Ⓑ | Ⓒ    Ⓓ |
| 9. | the solution to $h - \frac{6}{7} = \frac{2}{3}$ | the solution to $h - \frac{7}{9} = \frac{1}{6}$ |
| | Ⓐ    Ⓑ | Ⓒ    Ⓓ |

10. **Multiple Choice** In an outstanding season the Dodgers scored 41 runs in 8 games. Their opponents scored only 18 runs. Which equation would you use to determine the number of runs by which they outscored their opponents?

    Ⓐ $r - 41 = 18$ Ⓑ $r + 41 = 18$
    Ⓒ $r - 18 = 41$ Ⓓ $r + 18 = 41$

11. **Multiple Choice** The temperature at the beach was 98°F by 2:00 in the afternoon. Early in the evening the temperature was 69°F. What was the change in temperature?

    Ⓐ 31°F Ⓑ $-31$°F
    Ⓒ 29°F Ⓓ $-29$°F

# Standardized Test Practice
## 3.2 Solving Equations by Multiplying and Dividing

**TEST TAKING STRATEGY** Read the question again to be sure you answered it correctly.

1. **Multiple Choice** What reciprocal would be used to solve the equation $3p = -12$?
   - Ⓐ $\frac{1}{12}$
   - Ⓑ $\frac{1}{3}$
   - Ⓒ $\frac{3}{1}$
   - Ⓓ $-3$

2. **Multiple Choice** To solve the equation $-24 = -\frac{y}{6}$, which operation is used to isolate the variable $y$?
   - Ⓐ addition
   - Ⓑ division
   - Ⓒ subtraction
   - Ⓓ multiplication

3. **Multiple Choice** What is the solution to the equation $\frac{-7w}{5} = \frac{4}{10}$?
   - Ⓐ $w = -\frac{2}{7}$
   - Ⓑ $w = -\frac{14}{25}$
   - Ⓒ $w = -\frac{7}{2}$
   - Ⓓ $w = -\frac{25}{14}$

4. **Multiple Choice** Madeleine's family wants to make a 487.5-mile trip in $7\frac{1}{2}$ hours. What speed must they average for the trip to make it in time?
   - Ⓐ 55 miles per hour
   - Ⓑ 60 miles per hour
   - Ⓒ 65 miles per hour
   - Ⓓ 70 miles per hour

5. **Multiple Choice** What is the first step to solve the equation $18 = \frac{m}{-9}$?
   - Ⓐ Subtract 9 from both sides.
   - Ⓑ Add 9 to both sides.
   - Ⓒ Multiply both sides by $-9$.
   - Ⓓ Divide both sides by $-9$.

**Quantitative Comparison** In Exercises 6–8, choose the letter of the statement below that is true about the quantities in Columns I and II.

- **A** The number in Column I is greater.
- **B** The number in Column II is greater.
- **C** The two numbers are equal.
- **D** The relationship cannot be determined from the given information.

| | Column I | Column II |
|---|---|---|
| 6. | the value of $a$ | |
| | $-2a = -90$ | $\frac{a}{15} = 3$ |
| | Ⓐ  Ⓑ | Ⓒ  Ⓓ |
| 7. | the value of $c$ | |
| | $6c = -29$ | $15c = 10$ |
| | Ⓐ  Ⓑ | Ⓒ  Ⓓ |
| 8. | the value of $h$ | |
| | $-0.25 = \frac{h}{-5}$ | $0.325 = 0.325h$ |
| | Ⓐ  Ⓑ | Ⓒ  Ⓓ |

9. **Multiple Choice** One video game costs $29.95. How many games can Kathy buy with $179.70?
   - Ⓐ 5
   - Ⓑ 5.2
   - Ⓒ 6
   - Ⓓ 6.2

10. **Multiple Choice** Which equation best represents how to find the cost of one video tape if 8 tapes cost $10.80?
    - Ⓐ $8v = 10.8$
    - Ⓑ $10.8v = 8$
    - Ⓒ $\frac{v}{8} = 10.8$
    - Ⓓ $\frac{v}{10.8} = 8$

Algebra 1      Standardized Test Practice 3.2    17

NAME _____ CLASS _____ DATE _____

# Standardized Test Practice
## 3.3 Solving Two-Step Equations

**TEST TAKING STRATEGY**  Use an estimate to check to see that the answer is reasonable.

1. **Multiple Choice** The solution $-22.2$ is true for which equation?
   - Ⓐ $\dfrac{n}{2} + 7 = -4.1$
   - Ⓑ $2n + 7 = -4.1$
   - Ⓒ $\dfrac{n}{2} - 7 = 4.1$
   - Ⓓ $\dfrac{n}{2} - 2 = 4.12$

2. **Multiple Choice** Which is the solution to the equation $26 = -3g - 4$?
   - Ⓐ $g < 0$
   - Ⓑ $g > -15$
   - Ⓒ $g = -10$
   - Ⓓ all of the above

3. **Multiple Choice** What is the solution to the equation $-2\dfrac{1}{2} + \dfrac{w}{-2} = 2\dfrac{1}{3}$?
   - Ⓐ $w = -\dfrac{1}{10}$
   - Ⓑ $w = -\dfrac{1}{3}$
   - Ⓒ $w = -8\dfrac{5}{6}$
   - Ⓓ $w = -9\dfrac{2}{3}$

4. **Multiple Choice** Suzy bought 3 pairs of shorts and one pair of slacks. Without tax the total was $78.85. Which is the equation that would determine the price for each pair of shorts if the slacks cost $25.00?
   - Ⓐ $3x + 25 = 78.85$
   - Ⓑ $3x + 78.85 = 25$
   - Ⓒ $3(25) + x = 78.85$
   - Ⓓ $3(x + 25) = 78.85$

5. **Multiple Choice** Chrissy has 32 more customers than twice as many customers as when she first started her business. She now has 78 customers. How many customers did she have when she started her business?
   - Ⓐ 39 customers
   - Ⓑ 55 customers
   - Ⓒ 23 customers
   - Ⓓ 16 customers

**Quantitative Comparison** In Exercises 6–8, choose the letter of the statement below that is true about the quantities in Columns I and II.

- **A** The number in Column I is greater.
- **B** The number in Column II is greater.
- **C** The two numbers are equal.
- **D** The relationship cannot be determined from the given information.

| | Column I | Column II |
|---|---|---|
| 6. | \multicolumn{2}{c}{the value of $x$} |
| | $-3 = 5x + 4$ | $-8x + 48 = -36$ |
| | Ⓐ   Ⓑ | Ⓒ   Ⓓ |
| 7. | the value of $x$ in | the value of $y$ in |
| | $\dfrac{x}{-3} + 4.7 = 2.5$ | $-3y - 13 = -3$ |
| | Ⓐ   Ⓑ | Ⓒ   Ⓓ |
| 8. | \multicolumn{2}{c}{the value of $y$} |
| | $1\dfrac{1}{2} + y = -2\dfrac{1}{2}$ | $4.2 - y = 8.2$ |
| | Ⓐ   Ⓑ | Ⓒ   Ⓓ |

9. **Multiple Choice** The selling price of $179 for a cordless phone is $43 more than twice the wholesale cost. What is the wholesale cost of a cordless phone?
   - Ⓐ $111
   - Ⓑ $89.50
   - Ⓒ $68
   - Ⓓ $132.50

10. **Multiple Choice** What is the first step in solving the equation $3 = \dfrac{a}{3} - 7$?
    - Ⓐ Subtract 3 from both sides.
    - Ⓑ Divide both sides by 3.
    - Ⓒ Add 7 to both sides.
    - Ⓓ Subtract 7 from both sides.

NAME _____ CLASS _____ DATE _____

# Standardized Test Practice
## 3.4 Solving Multistep Equations

**TEST TAKING STRATEGY** Use context clues to help define an unknown word.

1. **Multiple Choice** What is the solution of the equation $8x - 10 = 6x - 12 + 4x$?
   - Ⓐ $x = -\frac{11}{9}$
   - Ⓑ $x = 1$
   - Ⓒ $x = -1$
   - Ⓓ $x = -11$

2. **Multiple Choice** What is the least common denominator in the equation $\frac{m}{4} + 2 = \frac{m}{3} + \frac{1}{2}$?
   - Ⓐ 2
   - Ⓑ 8
   - Ⓒ 24
   - Ⓓ 12

3. **Multiple Choice** Which equation has a solution that is negative?
   - Ⓐ $-5w + 2 = 15w - 10$
   - Ⓑ $2.5w + 1.3 = 1.3w - 2.3$
   - Ⓑ $-4w = -12$
   - Ⓑ $12 - 6w = 10 - 5w$

4. **Multiple Choice** Gary is studying for his fourth biology test. His first three test scores were 82, 93, and 87. Gary needs to know what he must score on this test in order to have an average of exactly 90. Which equation can be used to find the score Gary needs?
   - Ⓐ $\frac{x + 82 + 93 + 87}{4} = 90$
   - Ⓑ $4(x + 82 + 93 + 87) = 90$
   - Ⓒ $\frac{x}{4} + 82 + 93 + 87 = 90$
   - Ⓓ $x + 82 + 93 + 87 = \frac{90}{4}$

5. **Multiple Choice** What property is being illustrated below?
   $$\frac{x}{2} - 3 = 15 - \frac{2x}{3}$$
   $$\frac{x}{2} - 3 + \frac{2x}{3} = 15 - \frac{2x}{3} + \frac{2x}{3}$$
   - Ⓐ Property of Zero
   - Ⓑ Addition Property of Equality
   - Ⓒ Division Property of Equality
   - Ⓓ Distributive Property

*Quantitative Comparison* In Exercises 6–7, choose the letter of the statement below that is true about the quantities in Columns I and II.

**A** The number in Column I is greater.
**B** The number in Column II is greater.
**C** The two numbers are equal.
**D** The relationship cannot be determined from the given information.

| | Column I | | Column II |
|---|---|---|---|
| 6. | | the value of $x$ | |
| | $7 - 3x = 4x + 1$ | | $2x - 3 = \frac{x}{2} + 2$ |
| | Ⓐ Ⓑ | | Ⓒ Ⓓ |
| 7. | | the value of $x$ | |
| | $3x - 7 = 6x + 2$ | | $4x + 3 = 5x - 4$ |
| | Ⓐ Ⓑ | | Ⓒ Ⓓ |

8. **Multiple Choice** In the equation $\frac{2}{3}c + \frac{1}{2} = \frac{1}{5}c + \frac{3}{10}$, which is the best way to begin solving for $c$?
   - Ⓐ Add all of the numerators.
   - Ⓑ Subtract fractions.
   - Ⓒ Divide by LCD.
   - Ⓓ Multiply by LCD.

Algebra 1          Standardized Test Practice 3.4

NAME _____ CLASS _____ DATE _____

# Standardized Test Practice
## 3.5 Using the Distributive Property

**TEST TAKING STRATEGY** Restate the question and decide if you answered it completely.

1. **Multiple Choice** In the equation $5x - 3(2x + 4) = 17$, what is one of the values obtained after using the Distributive Property?
   - Ⓐ 7
   - Ⓑ −12
   - Ⓒ 12
   - Ⓓ −3

2. **Multiple Choice** Which correctly shows the use of the Distributive Property on the left side of $-3(1.5x + 3.2) = -1.3x$?
   - Ⓐ $-3.15x - 9.6$
   - Ⓑ $3.55x + 6.5$
   - Ⓒ $-4.5x - 9.6$
   - Ⓓ $-4.5x + 9.6$

3. **Multiple Choice** What is the solution to the equation $4(-3x - 5) = -2(5x + 3)$?
   - Ⓐ $x = -7$
   - Ⓑ $x = \dfrac{13}{11}$
   - Ⓒ $x = -1$
   - Ⓓ $x = 13$

4. **Multiple Choice** In order to find the perimeter of the rectangle, which equation can be used?

   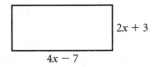

   - Ⓐ $P = (2x + 3) + (4x - 7)$
   - Ⓑ $P = 2(2x + 3) + (4x - 7)$
   - Ⓒ $P = (2x + 3) + 2(4x - 7)$
   - Ⓓ $P = 2(2x + 3) + 2(4x - 7)$

5. **Multiple Choice** The sum of the measures of the angles of triangle ABC is 180°. If the measures of angles A and B are each 27°, which equation can be used to find the measure of angle C?
   - Ⓐ $a + 27 = 180$
   - Ⓑ $2(27) - a = 180$
   - Ⓒ $2a + 27 = 180$
   - Ⓓ $2(27) + a = 180$

*Quantitative Comparison* In Exercises 6–9, choose the letter of the statement below that is true about the quantities in Columns I and II.

- **A** The number in Column I is greater.
- **B** The number in Column II is greater.
- **C** The two numbers are equal.
- **D** The relationship cannot be determined from the given information.

|   | Column I | Column II |
|---|----------|-----------|
| 6. | \multicolumn{2}{c}{the value of $x$} ||
|   | $-3 = 5(x + 4)$ | $-8(x + 6) = -36$ |
|   | Ⓐ  Ⓑ | Ⓒ  Ⓓ |
| 7. | the value of $y$ ||
|   | $3 - 2(y - 3) = 3$ | $3(y + 3) = 8$ |
|   | Ⓐ  Ⓑ | Ⓒ  Ⓓ |
| 8. | the value of $r$ ||
|   | $0.4 - 0.2r = r + 4$ | $2(r + 0.7) = 5.6$ |
|   | Ⓐ  Ⓑ | Ⓒ  Ⓓ |
| 9. | the perimeter of the rectangle ||

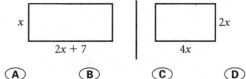

Ⓐ  Ⓑ  Ⓒ  Ⓓ

10. **Multiple Choice** In the equation $9k - 4(3k + 4) = 2k + (5k - 21)$, what step should be completed first in order to solve for $k$?
    - Ⓐ Combine like terms.
    - Ⓑ Add 21 to both sides.
    - Ⓒ Distribute −4.
    - Ⓓ Distribute $2k$.

NAME _____ CLASS _____ DATE _____

# Standardized Test Practice
## 3.6 Using Formulas and Literal Equations

**TEST TAKING STRATEGY** Keep in mind what you know and that you can make an inference to fill in missing information.

1. **Multiple Choice** Which describes an equation that involves two or more variables?
   - Ⓐ literal
   - Ⓑ formula
   - Ⓒ function
   - Ⓓ multi-equation

2. **Multiple Choice** If today's temperature is 50°F, and Kathie needs to record it in degrees Celsius, what value should she record? Use the formula $C = \frac{5}{9}(F - 32)$.
   - Ⓐ 10°C
   - Ⓑ 32.4°C
   - Ⓒ 20°C
   - Ⓓ −15°C

3. **Multiple Choice** The formula for density is $D = \frac{m}{V}$, where $m$ = mass and $V$ = volume. If you know the mass and density of a substance, how can you find the volume?
   - Ⓐ $V = D \times m$
   - Ⓑ $V = \frac{m}{D}$
   - Ⓒ $V = \frac{D}{m}$
   - Ⓓ $V = D + m$

4. **Multiple Choice** Which describes the first step in solving the equation $ax + b = c$ for $b$?
   - Ⓐ Subtract $b$ from both sides.
   - Ⓑ Divide both sides by $a$.
   - Ⓒ Subtract $ax$ from both sides.
   - Ⓓ Divide both sides by $x$.

5. **Multiple Choice** Use the formula $C = 2\pi r$ to find the radius, $r$, of a circle when the circumference, $C$, is 70 meters.
   - Ⓐ 439.60 meters
   - Ⓑ 44.56 meters
   - Ⓒ 11.15 meters
   - Ⓓ 22.29 meters

**Quantitative Comparison** In Exercises 6–9, choose the letter of the statement below that is true about the quantities in Columns I and II.

- **A** The number in Column I is greater.
- **B** The number in Column II is greater.
- **C** The two numbers are equal.
- **D** The relationship cannot be determined from the given information.

| | Column I | Column II |
|---|---|---|
| 6. | \multicolumn{2}{c}{$P = 2l + 2w$} |
| | $l$ | $w$ |
| | Ⓐ   Ⓑ | Ⓒ   Ⓓ |
| 7. | \multicolumn{2}{c}{$C = \frac{5}{9}(F - 32)$} |
| | the value of 95°F in degree Celsius | 37°C |
| | Ⓐ   Ⓑ | Ⓒ   Ⓓ |
| 8. | | $C = 2\pi r$; $C = 45$ centimeters |
| | $r = 8$ centimeters | $r$ |
| | Ⓐ   Ⓑ | Ⓒ   Ⓓ |
| 9. | the length of a rectangle whose perimeter is 48 inches, and whose length is 3 inches more than twice the width | the length of a rectangle whose perimeter is 56 inches, and whose length is 3 times the width |
| | Ⓐ   Ⓑ | Ⓒ   Ⓓ |

Algebra 1

# SAT/ACT Chapter Test
## Chapter 3 Equations

**TEST TAKING STRATEGY** Usually first instincts are correct. Only change your answer if you know for certain another choice is better.

1. **Multiple Choice** What is the first step in solving the equation $x + 78 = 135$?
   - A Add 78 to both sides.
   - B Subtract 78 from both sides.
   - C Add 135 to both sides.
   - D Subtract 135 from both sides.

2. **Multiple Choice** What is the solution to the equation $3(x + 2) = 9$?
   - A $x = 1$
   - B $x = 10$
   - C $x = 5$
   - D $x = 25$

3. **Multiple Choice** Jolanda bought 6 chairs and one table. Without tax the total was $154.50. Which equation can be used to determine the price for each chair, if the table cost $38.00?
   - A $6x - 38 = 154.50$
   - B $6x + 154.50 = 38.00$
   - C $6x + 38 = 154.50$
   - D $6(x + 25) = 154.50$

4. **Multiple Choice** The sum of the measures of the angles of triangle EFG is 180°. If the measures of angles E and G are each 54°, which equation can be used to find the measure of angle F?
   - A $f + 54 = 180$
   - B $2(54) - f = 180$
   - C $2(54) + f = 180$
   - D $2f + 54 = 180$

5. **Multiple Choice** Use the formula $C = 2\pi r$ to find the radius, $r$, when the circumference, $C$, of a circle is 45 centimeters.
   - A 7.17 centimeters
   - B 14.33 centimeters
   - C 8.26 centimeters
   - D 22.25 centimeters

6. **Multiple Choice** Which shows the equation $ax + \dfrac{b}{2} = c$ in terms of $b$?
   - A $b = ax - 2c$
   - B $b = 2(c - ax)$
   - C $b = \dfrac{c - ax}{2}$
   - D $b = 2(ax - c)$

**Quantitative Comparison** In Exercises 7–9, choose the letter of the statement below that is true about the quantities in Columns I and II.

A The number in Column I is greater.
B The number in Column II is greater.
C The two numbers are equal.
D The relationship cannot be determined from the given information.

| | Column I | Column II |
|---|---|---|
| 7. | the value of $x$ | |
| | $x + 17 = -4$ | $1.3 = \dfrac{x}{-5}$ |
| | A   B | C   D |
| 8. | the value of $x$ | |
| | $5x = 0$ | $2(x - 3) = x - 6$ |
| | A   B | C   D |
| 9. | the value of $m$ | |
| | $11 = 3 - \dfrac{m}{4}$ | $-3m = 5(m - 8)$ |
| | A   B | C   D |

10. **Multiple Choice** If the temperature of a liquid is 131°F, what is its equivalence in degrees Celsius? Use the formula $C = \dfrac{5}{9}(F - 32)$.
    - A 55°C
    - B 55.5°C
    - C 90.5°C
    - D 45°C

NAME _____ CLASS _____ DATE _____

# Standardized Test Practice
## 4.1 Using Proportional Reasoning

**TEST TAKING STRATEGY** Use number sense to eliminate unreasonable choices.

1. **Multiple Choice** What value of $n$ makes the following a proportion?
$$\frac{9}{15} = \frac{n}{60}$$
   - Ⓐ 4
   - Ⓑ 36
   - Ⓒ 45
   - Ⓓ 27

2. **Multiple Choice** What statement is equivalent to $\frac{18}{24} = \frac{6}{x}$?
   - Ⓐ $18x = 24 \cdot 6$
   - Ⓑ $18 \cdot 6 = 24 \cdot x$
   - Ⓒ $24 \cdot 6 \cdot x = 18$
   - Ⓓ $x = 24 \cdot 6 \cdot 18$

3. **Multiple Choice** What values for $t$ and $s$ do *not* make a proportion, $\frac{t}{s} = \frac{72}{24}$?
   - Ⓐ $t = 9, s = 3$
   - Ⓑ $t = 12, s = 6$
   - Ⓒ $t = 24, s = 8$
   - Ⓓ $t = 36, s = 12$

4. **Multiple Choice** Solve $\frac{18}{y} = \frac{2}{5}$ for $y$.
   - Ⓐ 9
   - Ⓑ 45
   - Ⓒ 18
   - Ⓓ 25

5. **Multiple Choice** In the proportion $\frac{6}{21} = \frac{x}{28}$, the value of $x$ is:
   - Ⓐ less than 6
   - Ⓑ 7 more than 21
   - Ⓒ more than 28
   - Ⓓ greater than 6

6. **Multiple Choice** For every 72 students there are 16 females. If there are 4 females, about how many students are there?
   - Ⓐ 5
   - Ⓑ 7
   - Ⓒ 21
   - Ⓓ 18

**Quantitative Comparison** In Exercises 7–9, choose the letter of the statement below that is true about the quantities in Columns I and II.

- **A** The number in Column I is greater.
- **B** The number in Column II is greater.
- **C** The two numbers are equal.
- **D** The relationship cannot be determined from the given information.

| | Column I | Column II |
|---|---|---|
| 7. | the value of $x$ in $\frac{6}{5} = \frac{24}{x}$ | the value of $y$ in $\frac{6}{y} = \frac{3}{7}$ |
| | Ⓐ  Ⓑ | Ⓒ  Ⓓ |
| 8. | the value of $a$ in $\frac{19}{38} = \frac{17}{a}$ | the value of $c$ in $\frac{19}{3} = \frac{c}{15}$ |
| | Ⓐ  Ⓑ | Ⓒ  Ⓓ |
| 9. | the value of $t$ in $\frac{7}{8} = \frac{49}{t}$ | the value of $r$ in $\frac{3}{14} = \frac{12}{r}$ |
| | Ⓐ  Ⓑ | Ⓒ  Ⓓ |

10. **Multiple Choice** If CDs are on sale at 3 for $20, how much will 9 CDs cost?
    - Ⓐ $40
    - Ⓑ $50
    - Ⓒ $60
    - Ⓓ $90

11. **Multiple Choice** A 25-foot tree casts a 15-foot shadow. How long is the shadow cast by a 5-foot pole at this time?
    - Ⓐ 5 feet
    - Ⓑ 10 feet
    - Ⓒ 15 feet
    - Ⓓ 3 feet

Algebra 1 — Standardized Test Practice 4.1

NAME _____ CLASS _____ DATE _____

# Standardized Test Practice
## 4.2 Percent Problems

**TEST TAKING STRATEGY** Avoid changing your answer unless you know another one is better.

**1. Multiple Choice** What is 30% of 500?
- Ⓐ 15
- Ⓑ 150
- Ⓒ 1,500
- Ⓓ 15,000

**2. Multiple Choice** What percent of 75 is 15?
- Ⓐ 5%
- Ⓑ 25%
- Ⓒ 2%
- Ⓓ 20%

**3. Multiple Choice** 12 is 30% of what number?
- Ⓐ 36
- Ⓑ 400
- Ⓒ 40
- Ⓓ 3,600

**4. Multiple Choice** What is 125% of 72?
- Ⓐ 80
- Ⓑ 90
- Ⓒ 180
- Ⓓ 18

**5. Multiple Choice** The winning candidate in an election received 56% of the votes cast. The winner received 42,000 votes. How many votes were cast?
- Ⓐ 23,520
- Ⓑ 255,200
- Ⓒ 75,000
- Ⓓ 750,000

**6. Multiple Choice** 236 is 118% of what number?
- Ⓐ 200
- Ⓑ 278.48
- Ⓒ 218
- Ⓓ 136

**7. Multiple Choice** The decimal ____ is equivalent to 85%.
- Ⓐ 0.85
- Ⓑ 0.085
- Ⓒ 8.50
- Ⓓ 0.805

**Quantitative Comparison** In Exercises 8–11, choose the letter of the statement below that is true about the quantities in Columns I and II.

- **A** The number in Column I is greater.
- **B** The number in Column II is greater.
- **C** The two numbers are equal.
- **D** The relationship cannot be determined from the given information.

|     | Column I | Column II |
|-----|----------|-----------|
| 8.  | 20% of 45 | 50% of 18 |
|     | Ⓐ Ⓑ | Ⓒ Ⓓ |
| 9.  | $\frac{3}{5}$ of $x$ if $x > 0$ | 52% of $x$ if $x > 0$ |
|     | Ⓐ Ⓑ | Ⓒ Ⓓ |
| 10. | 25% of 76 | 125% of 16 |
|     | Ⓐ Ⓑ | Ⓒ Ⓓ |
| 11. | 42% of $x$ | 13% of $y$ |
|     | Ⓐ Ⓑ | Ⓒ Ⓓ |

**12. Multiple Choice** 84 is what percent of 56?
- Ⓐ $66\frac{2}{3}$%
- Ⓑ 126%
- Ⓒ 150%
- Ⓓ 75%

**13. Multiple Choice** A student answered 15 questions correctly on a quiz, for a score of 75%. How many questions were on the quiz?
- Ⓐ 12
- Ⓑ 18
- Ⓒ 20
- Ⓓ 25

# Standardized Test Practice
## 4.3 Introduction to Probability

**TEST TAKING STRATEGY** Write down all known formulas and rules on a scrap piece of paper.

1. **Multiple Choice** Two number cubes are rolled 100 times. An odd number appears on both cubes 23 times. Which value shows the experimental probability of this outcome?
   - Ⓐ 46%
   - Ⓑ 23%
   - Ⓒ 50%
   - Ⓓ 77%

2. **Multiple Choice** Two coins are flipped 200 times. The experimental probability of this outcome is 27%. How many times did two heads appear?
   - Ⓐ 28
   - Ⓑ 46
   - Ⓒ 32
   - Ⓓ 54

3. **Multiple Choice** Two number cubes are rolled 100 times. How many times did a sum of 6 appear if the experimental probability of this outcome is 4%?
   - Ⓐ 25
   - Ⓑ 4
   - Ⓒ 10
   - Ⓓ 1

4. **Multiple Choice** Two coins are flipped 20 times. Both coins are tails 6 times. What is the experimental probability of this outcome?
   - Ⓐ 60%
   - Ⓑ 28%
   - Ⓒ 30%
   - Ⓓ 25%

5. **Multiple Choice** Two coins are flipped 100 times. One head and one tail appear 48 times. The experimental probability of this outcome is best described as
   - Ⓐ less than 50%.
   - Ⓑ more than 50%.
   - Ⓒ almost 100%.
   - Ⓓ less than 25%.

**Quantitative Comparison** In Exercises 6–8, choose the letter of the statement below that is true about the quantities in Columns I and II.

A The number in Column I is greater.
B The number in Column II is greater.
C The two numbers are equal.
D The relationship cannot be determined from the given information.

| Column I | Column II |
|---|---|
| 6. The experimental probability of: | |
| 12 favorable outcomes out of 50 trials | 5 favorable outcomes out of 20 trials |
| Ⓐ Ⓑ | Ⓒ Ⓓ |
| 7. The experimental probability of: | |
| 18 favorable outcomes out of 100 trials | 35 favorable outcomes out of 200 trials |
| Ⓐ Ⓑ | Ⓒ Ⓓ |
| 8. $\frac{64}{200}$ | 32% |
| Ⓐ Ⓑ | Ⓒ Ⓓ |

9. **Multiple Choice** Aleta flipped 2 coins 20 times and got 2 heads 6 times. Juan flipped 2 coins 20 times and got 2 heads 4 times. If Aleta's and Juan's results are combined, what is the experimental probability of getting 2 heads?
   - Ⓐ 50%
   - Ⓑ 30%
   - Ⓒ 20%
   - Ⓓ 25%

NAME _____ CLASS _____ DATE _____

# Standardized Test Practice
## 4.4 Measures of Central Tendency

**TEST TAKING STRATEGY** Use context to help define an unknown word.

1. **Multiple Choice** Which value best describes the range of these temperatures?

    75°, 92°, 78°, 72°, 73°, 90°, 70°

    Ⓐ 15°  Ⓑ 22°
    Ⓒ 17°  Ⓓ 20°

2. **Multiple Choice** What is the median in the following set of data?

    58 kg, 54 kg, 55 kg, 61 kg,
    67 kg, 52 kg, 60 kg

    Ⓐ 58.1 kg  Ⓑ 15 kg
    Ⓒ 61 kg    Ⓓ 58 kg

3. **Multiple Choice** Shea exercised 8 days in a row. Her daily exercise time, in hours, varied as follows:

    $\frac{1}{2}, 1\frac{1}{2}, 1, \frac{3}{4}, \frac{1}{2}, 1, \frac{3}{4}, \frac{1}{2}$

    What is the mode of the data?

    Ⓐ $\frac{1}{2}$  Ⓑ $\frac{3}{4}$
    Ⓒ 1            Ⓓ $1\frac{1}{2}$

4. **Multiple Choice** If Ms. Lind wants to know the average height of a boy in her class, what measure of central tendency should she use?

    Ⓐ mode   Ⓑ range
    Ⓒ mean   Ⓓ median

5. **Multiple Choice** What is the mean of the following test scores?

    77%, 65%, 88%, 90%, 95%, 85%

    Ⓐ $86.\overline{5}$  Ⓑ 30
    Ⓒ $83.\overline{3}$  Ⓓ 100

**Quantitative Comparison** In Exercises 6–8, choose the letter of the statement below that is true about the quantities in Columns I and II.

A The number in Column I is greater.
B The number in Column II is greater.
C The two numbers are equal.
D The relationship cannot be determined from the given information.

|    | Column I | Column II |
|----|----------|-----------|
| 6. | 12, 22, 18, 25, 16 | |
|    | mean | median |
|    | Ⓐ  Ⓑ  Ⓒ  Ⓓ | |
| 7. | 101, 106, 101, 108, 110 | |
|    | range | mode |
|    | Ⓐ  Ⓑ  Ⓒ  Ⓓ | |
| 8. | 45, 55, 55, 48, 39 | |
|    | mean | mode |
|    | Ⓐ  Ⓑ  Ⓒ  Ⓓ | |

9. **Multiple Choice** A batter has 18 hits in 60 times at bat. How many hits does he need in his next 60 times at bat in order to have a batting average of 0.333?

    Ⓐ 18  Ⓑ 40
    Ⓒ 20  Ⓓ 22

10. **Multiple Choice** Which set of data has the same mean, median, and mode?

    Ⓐ 5, 7, 7, 7, 9
    Ⓑ 1, 2, 2, 3, 5
    Ⓒ 6, 8, 8, 8, 9
    Ⓓ 6, 7, 7, 7, 10

# Standardized Test Practice
## 4.5 Graphing Data

**TEST TAKING STRATEGY** Read the graphs and diagrams for useful information.

A wildlife conservation organization had $80 million in revenue last year. The circle graph shows the sources of the revenue. Use the graph for Exercises 1–3.

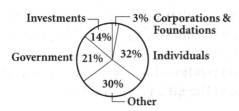

1. **Multiple Choice** How much of the revenue came from individuals?

   Ⓐ $2.56 million  Ⓑ $25.6 million
   Ⓒ $256 million  Ⓓ $256,000

2. **Multiple Choice** How much more of the revenue came from government funding than from investments?

   Ⓐ $56 million  Ⓑ $11.2 million
   Ⓒ $5.6 million  Ⓓ $16.8 million

3. **Multiple Choice** Funds from corporations and foundations accounted for how much of the revenue?

   Ⓐ $24 million  Ⓑ $240,000
   Ⓒ $24,000  Ⓓ $2.4 million

4. **Multiple Choice** Which bar graph shows a year-to-year increase over the 4-year period?

   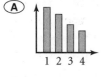 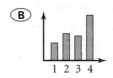
    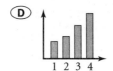

*Quantitative Comparison* In Exercises 5–7, choose the letter of the statement below that is true about the quantities in Columns I and II.

A The number in Column I is greater.
B The number in Column II is greater.
C The two numbers are equal.
D The relationship cannot be determined from the given information.

| | Column I | Column II |
|---|---|---|
| 5. | the number of degrees in a pie-chart sector that represents 22% | the number of degrees in a pie-chart sector that represents 19.5% |
| | Ⓐ  Ⓑ | Ⓒ  Ⓓ |
| 6. | in a bar graph, the height of the bar that represents $7,500,000 annual sales in 1999 | $9,250,000 annual sales in 2000 |
| | Ⓐ  Ⓑ | Ⓒ  Ⓓ |
| 7. | in a pie chart, the number of degrees that represent 42 voters out of a sample of 200 | in the pie chart, the number of degrees that represent 33% of the voters |
| | Ⓐ  Ⓑ | Ⓒ  Ⓓ |

8. **Multiple Choice** In the line graph, which 6-hour period shows the greatest increase in temperature?

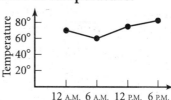

   Ⓐ 12 A.M. to 6 A.M.  Ⓑ 6 A.M. to 12 P.M.
   Ⓒ 12 P.M. to 6 P.M.  Ⓓ 6 A.M. to 6 P.M.

# Standardized Test Practice
## 4.6 Other Data Displays

**TEST TAKING STRATEGY** Use the test information as a tool to help you answer other questions.

The table shows the lengths in centimeters of 20 blue-tailed skinks, small reptiles found in the tropical forests of Pacific islands. Use this information for Exercises 1–4.

| 12.2 | 13.5 | 14.6 | 15.3 | 15.9 | 16.4 | 16.7 |
| 17.2 | 18.4 | 18.9 | 20.1 | 21.3 | 21.3 | 22.4 |
| 23.3 | 23.7 | 24.5 | 26.4 | 27.3 | 27.9 |

1. **Multiple Choice** In a stem-and-leaf plot of the data, what set of whole numbers should be used as the stems?
   - Ⓐ 1–20
   - Ⓑ 10–30
   - Ⓒ 15–30
   - Ⓓ 12–27

2. **Multiple Choice** In a stem-and-leaf plot, how many stems would have 2 leaves?
   - Ⓐ 6
   - Ⓑ 5
   - Ⓒ 2
   - Ⓓ 4

3. **Multiple Choice** In a histogram of the data, the intervals 10.0–14.9, 15.0–19.9, 20.0–24.9, and 25.0–29.9 are used. What is the frequency of each interval, respectively?
   - Ⓐ 4, 6, 7, 3
   - Ⓑ 3, 7, 6, 4
   - Ⓒ 3, 7, 7, 3
   - Ⓓ 3, 3, 7, 8

4. **Multiple Choice** Which box-and-whisker plot best represents the data?

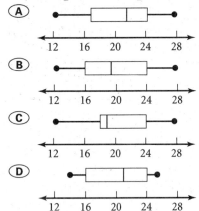

*Quantitative Comparison* In Exercises 5–7, choose the letter of the statement below that is true about the quantities in Columns I and II.

A The number in Column I is greater.
B The number in Column II is greater.
C The two numbers are equal.
D The relationship cannot be determined from the given information.

| | Column I | Column II |
|---|---|---|
| 5. | the highest number in a set of data | the upper quartile in the same set of data |
| | Ⓐ  Ⓑ | Ⓒ  Ⓓ |
| 6. | the lower quartile in a set of data | the median in the same set of data |
| | Ⓐ  Ⓑ | Ⓒ  Ⓓ |
| 7. | the median in a set of data | the median in a different set of data |
| | Ⓐ  Ⓑ | Ⓒ  Ⓓ |

8. **Multiple Choice** Which kind of graph is similar to a histogram?
   - Ⓐ a line graph
   - Ⓑ a bar graph
   - Ⓒ a circle graph
   - Ⓓ a pictograph

9. **Multiple Choice** In a stem-and-leaf plot for a data set that consists of numbers between 151 and 192, what numbers might be used as stems?
   - Ⓐ 15–19
   - Ⓑ 150–190
   - Ⓒ 1–10
   - Ⓓ 151–192

# SAT/ACT Chapter Test

## Chapter 4  Proportional Reasoning and Statistics

**TEST TAKING STRATEGY**  Use an estimate to check to see that your answer is reasonable.

1. **Multiple Choice**  What value of $n$ makes the expression a proportion?

   $$\frac{12}{n} = \frac{66}{44}$$

   Ⓐ 10    Ⓑ 8
   Ⓒ 9     Ⓓ 6

2. **Multiple Choice**  105 is to 49 as $x$ is to 7. What is the value of $x$?

   Ⓐ 735   Ⓑ 7
   Ⓒ 15    Ⓓ 14

3. **Multiple Choice**  Greta only owes 85% of the $140 she borrowed from Sal. How much does she still owe Sal?

   Ⓐ $119    Ⓑ $1190
   Ⓒ $11.90  Ⓓ $1.19

4. **Multiple Choice**  78 is 150% of what number?

   Ⓐ 117   Ⓑ 11.7
   Ⓒ 50    Ⓓ 52

5. **Multiple Choice**  An antique market marked up a $60 table to $600. What is the percent of mark up?

   Ⓐ 100%   Ⓑ 1,000%
   Ⓒ 10%    Ⓓ 600%

6. **Multiple Choice**  Two coins are flipped 20 times. Both coins come up heads 4 times. What best describes the experimental probability of this outcome?

   Ⓐ 20%
   Ⓑ 25%
   Ⓒ 50%
   Ⓓ 40%

**Quantitative Comparison**  In Exercises 7–9, choose the letter of the statement below that is true about the quantities in Columns I and II.

A  The number in Column I is greater.
B  The number in Column II is greater.
C  The two numbers are equal.
D  The relationship cannot be determined from the given information.

| | Column I | Column II | |
|---|---|---|---|
| 7. | 18% of 75 | 10% of 135 | |
| | Ⓐ Ⓑ | Ⓒ Ⓓ | |
| 8. | the mean of a set of data | the median of a set of data | |
| | Ⓐ Ⓑ | Ⓒ Ⓓ | |
| 9. | the number of degrees in a pie-chart sector that represents 15% | the number of degrees in a pie-chart sector that represents 18% | |
| | Ⓐ Ⓑ | Ⓒ Ⓓ | |

Use the box-and-whisker plot for Exercises 10 and 11.

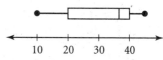

10. **Multiple Choice**  What is the median of the data?

    Ⓐ 30   Ⓑ 40
    Ⓒ 32   Ⓓ 20

11. **Multiple Choice**  What is the upper quartile of the data?

    Ⓐ 45   Ⓑ 50
    Ⓒ 40   Ⓓ 32

NAME _____ CLASS _____ DATE _____

# Standardized Test Practice
## 5.1 Linear Functions and Graphs

**TEST TAKING STRATEGY** Eliminate the obvious distracters from the answer choices.

1. **Multiple Choice** Which relation defines a function?
   - Ⓐ $(6, 3), (-4, -2), (6, 0)$
   - Ⓑ $(2, 5), (4, 8), (8, 5)$
   - Ⓒ $(-1, 2), (0, 4), (-1, 3)$
   - Ⓓ $(6, 4), (6, 3), (6, 2)$

2. **Multiple Choice** What number completes (4, ?) so that it is a solution of $2x + y = 10$?
   - Ⓐ $-2$
   - Ⓑ $18$
   - Ⓒ $2$
   - Ⓓ $6$

3. **Multiple Choice** What number completes (?, $-1$) so that it is a solution of $3x + 2y = 10$?
   - Ⓐ $2\frac{2}{3}$
   - Ⓑ $8$
   - Ⓒ $12$
   - Ⓓ $4$

Use the graph of a linear function for Exercises 4 and 5.

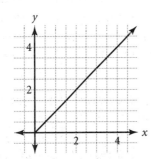

4. **Multiple Choice** Which ordered pair is in the function?
   - Ⓐ $(4, 4)$
   - Ⓑ $(2, 1.5)$
   - Ⓒ $(4, 3)$
   - Ⓓ $(6, 4.5)$

5. **Multiple Choice** What is an equation for the function?
   - Ⓐ $y = x$
   - Ⓑ $y = 0.75x$
   - Ⓒ $y = 1.5x$
   - Ⓓ $y = 2x$

6. **Multiple Choice** Which point lies on the graph of $x - 3y = 8$?
   - Ⓐ $(1, 3)$
   - Ⓑ $(1, -3)$
   - Ⓒ $(-1, 3)$
   - Ⓓ $(-1, -3)$

**Quantitative Comparison** In Exercises 7–9, choose the letter of the statement below that is true about the quantities in Columns I and II.

A  The number in Column I is greater.
B  The number in Column II is greater.
C  The two numbers are equal.
D  The relationship cannot be determined from the given information.

| | Column I | Column II |
|---|---|---|
| 7. | the value of $y$ in $(4, y)$ if $2x - y = 7$ | the value of $x$ in $(x, -5)$ if $2x - y = 7$ |
| | Ⓐ  Ⓑ | Ⓒ  Ⓓ |
| 8. | the value of $x$ in $(x, 2)$ if $y = 2$ | the value of $y$ in $(2, y)$ if $x = 2$ |
| | Ⓐ  Ⓑ | Ⓒ  Ⓓ |
| 9. | the value of $x$ in $(x, -3)$ if $x = 2y$ | the value of $y$ in $(0, y)$ if $2x - y = 5$ |
| | Ⓐ  Ⓑ | Ⓒ  Ⓓ |

10. **Multiple Choice** What is a linear equation for this function?

| $x$ | 0 | 1 | 2 | 3 |
|---|---|---|---|---|
| $y$ | $-3$ | 1 | 5 | 9 |

   - Ⓐ $y = 4x - 3$
   - Ⓑ $y = 3x - 4$
   - Ⓒ $y = 4x + 3$
   - Ⓓ $y = -3x + 4$

NAME _____ CLASS _____ DATE _____

# Standardized Test Practice
## 5.2 Defining Slope

**TEST TAKING STRATEGY** Restate each question to verify that you answered it correctly.

1. **Multiple Choice** Which kind of line has negative slope?
   - Ⓐ a line that slants down from left to right
   - Ⓑ a horizontal line
   - Ⓒ a vertical line
   - Ⓓ a line that slants up from left to right

2. **Multiple Choice** What is the slope of the line containing $(-4, 2)$ and $(3, 0)$?
   - Ⓐ $-\frac{2}{7}$
   - Ⓑ $3\frac{1}{2}$
   - Ⓒ $-3\frac{1}{2}$
   - Ⓓ $\frac{2}{7}$

Use the graph for Exercises 3–5.

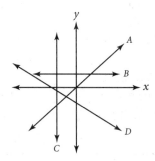

3. **Multiple Choice** Which line has slope 0?
   - Ⓐ line A
   - Ⓑ line B
   - Ⓒ line C
   - Ⓓ line D

4. **Multiple Choice** Which line has negative slope?
   - Ⓐ line A
   - Ⓑ line B
   - Ⓒ line C
   - Ⓓ line D

5. **Multiple Choice** For which line is the slope undefined?
   - Ⓐ line A
   - Ⓑ line B
   - Ⓒ line C
   - Ⓓ line D

**Quantitative Comparison** In Exercises 6–8, choose the letter of the statement below that is true about the quantities in Columns I and II.

- **A** The number in Column I is greater.
- **B** The number in Column II is greater.
- **C** The two numbers are equal.
- **D** The relationship cannot be determined from the given information.

| | Column I | Column II |
|---|---|---|
| 6. | the slope of the line containing $(4, -2)$ and $(6, -2)$ | the slope of the line containing $(-2, -4)$ and $(0, -2)$ |
| | Ⓐ  Ⓑ | Ⓒ  Ⓓ |
| 7. | the slope of a line that contains the origin | the slope of a line that does not contain the origin |
| | Ⓐ  Ⓑ | Ⓒ  Ⓓ |
| 8. | the slope of the line containing points $(0, 1)$ and $(2, 4)$ | the slope of the line containing points $(-1, -2)$ and $(3, 4)$ |
| | Ⓐ  Ⓑ | Ⓒ  Ⓓ |

9. **Multiple Choice** The foot of a ramp is 30 feet from the base of a building. The top of the ramp is 2 feet above the ground on the side of the building. What is the slope of the ramp?
   - Ⓐ $-\frac{2}{15}$
   - Ⓑ $\frac{1}{15}$
   - Ⓒ $15$
   - Ⓓ $-15$

10. **Multiple Choice** A slope of 3 is a line defined with the following characteristics:
    - Ⓐ rise: 1; run: 3
    - Ⓑ rise: $-3$; run: 1
    - Ⓒ rise: $-3$; run: $-1$
    - Ⓓ rise: 3; run: $-1$

NAME _____ CLASS _____ DATE _____

# Standardized Test Practice
## 5.3 Rates of Change and Direct Variation

**TEST TAKING STRATEGY** You can make an inference to fill in unknown information.

In Exercises 1–4, $y$ varies directly as $x$.

1. **Multiple Choice** If $y = 8$ when $x = 4$, what is the constant of variation?
   - Ⓐ 0.5
   - Ⓑ 2
   - Ⓒ 32
   - Ⓓ 0.25

2. **Multiple Choice** If $y = -1.2$ when $x = 2.4$, which equation shows a direct variation?
   - Ⓐ $y = 0.5x$
   - Ⓑ $y = -2x$
   - Ⓒ $y = -0.5x$
   - Ⓓ $y = 2x$

3. **Multiple Choice** If $y = 4$ when $x = 3$, what is $y$ when $x = 9$?
   - Ⓐ 6.75
   - Ⓑ 0.75
   - Ⓒ 12
   - Ⓓ 27

4. **Multiple Choice** If $y = 15$ and $k = \frac{1}{3}$, what does $x$ equal?
   - Ⓐ 45
   - Ⓑ 5
   - Ⓒ 15
   - Ⓓ 3

5. **Multiple Choice** Which graph best represents a direct variation?

   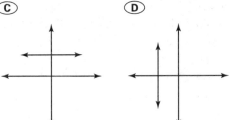

*Quantitative Comparison* In Exercises 6–8, choose the letter of the statement below that is true about the quantities in Columns I and II.

A The number in Column I is greater.
B The number in Column II is greater.
C The two numbers are equal.
D The relationship cannot be determined from the given information.

| Column I | Column II |
|---|---|
| 6. a rate of change of 60 miles per hour | a rate of change of 1 mile per hour |
| Ⓐ  Ⓑ | Ⓒ  Ⓓ |
| 7. the value of $k$ in $y = kx$ if $y = 11.5$ when $x = 2.3$ | the value of $k$ in $y = kx$ if $y = 9.6$ when $x = 2$ |
| Ⓐ  Ⓑ | Ⓒ  Ⓓ |
| 8. the slope of a line containing $(0, 0)$ and $(5, 20)$ | the value of $k$ in $y = kx$ if $y = 0$ when $x = 0$ |
| Ⓐ  Ⓑ | Ⓒ  Ⓓ |

9. **Multiple Choice** In the formula $A = lw$, for a given length, the area of a rectangle varies directly with the width. If $A = 21$ when $l = 7$, what is $w$?
   - Ⓐ $\frac{1}{3}$
   - Ⓑ 3
   - Ⓒ 147
   - Ⓓ $-14$

10. **Multiple Choice** Which is the direct-variation equation for converting meters, $m$, to kilometers, $k$?
    - Ⓐ $k = 0.001m$
    - Ⓑ $k = 0.01m$
    - Ⓒ $k = 1000m$
    - Ⓓ $-k = 0.1m$

NAME _____ CLASS _____ DATE _____

# Standardized Test Practice
## 5.4 The Slope-Intercept Form

**TEST TAKING STRATEGY** Look at each answer choice before choosing one.

1. **Multiple Choice** What are the $x$- and $y$-intercepts of $y = 4x + 2$?

    Ⓐ $x$-intercept: 0.5; $y$-intercept: 2
    Ⓑ $x$-intercept: 2; $y$-intercept: 0.5
    Ⓒ $x$-intercept: 4; $y$-intercept: 2
    Ⓓ $x$-intercept: $-0.5$; $y$-intercept: 2

2. **Multiple Choice** At which points does the line $y = -3.5x + 10.5$ intersect the axes?

    Ⓐ $(3, 0)$ and $(0, 10.5)$
    Ⓑ $(-3, 0)$ and $(0, 10.5)$
    Ⓒ $(-3.5, 0)$ and $(0, -10.5)$
    Ⓓ $(-3, 0)$ and $(0, -10.5)$

3. **Multiple Choice** What is an equation of the line containing points $(-2, 0)$ and $(0, -2)$?

    Ⓐ $y = -x - 2$  Ⓑ $y = -2x - 2$
    Ⓒ $y = x - 2$  Ⓓ $y = -x + 2$

4. **Multiple Choice** What is an equation of the line containing $(-3, 5)$ if it also contains the origin?

    Ⓐ $y = -\frac{3}{5}x$  Ⓑ $y = -\frac{5}{3}x + 5$
    Ⓒ $y = \frac{-5}{3}x$  Ⓓ $y = -\frac{5}{3}x - 3$

5. **Multiple Choice** What is an equation of the line containing $(-2, 6)$ and $(1, -6)$?

    Ⓐ $y = -4x + 14$  Ⓑ $y = -4x - 2$
    Ⓒ $y = 4x + 2$  Ⓓ $y = 4x - 14$

6. **Multiple Choice** Which line does *not* have an $x$-intercept?

    Ⓐ $y = 2x$  Ⓑ $y = -x$
    Ⓒ $x = -5$  Ⓓ $y = 3$

**Quantitative Comparison** In Exercises 7–9, choose the letter of the statement below that is true about the quantities in Columns I and II.

A The number in Column I is greater.
B The number in Column II is greater.
C The two numbers are equal.
D The relationship cannot be determined from the given information.

| | Column I | Column II | |
|---|---|---|---|
| 7. | the slope of $y = x + 8$ | the slope of $y = -0.5x - 2$ | |
| | Ⓐ Ⓑ | Ⓒ Ⓓ | |
| 8. | the $y$-intercept of $y = -4x + 5$ | the $y$-intercept of $y = 8x + 5$ | |
| | Ⓐ Ⓑ | Ⓒ Ⓓ | |
| 9. | the $x$-intercept of $y = x + 3$ | the slope of $y = -\frac{2}{3}x + 2$ | |
| | Ⓐ Ⓑ | Ⓒ Ⓓ | |

10. **Multiple Choice** What is the equation of the line shown?

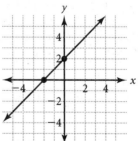

Ⓐ $y = -x - 2$  Ⓑ $y = x - 2$
Ⓒ $y = x + 2$  Ⓓ $y = -x + 2$

11. **Multiple Choice** Which line does *not* have a $y$-intercept?

    Ⓐ $y = -2x$  Ⓑ $y = 0$
    Ⓒ $x = 7$  Ⓓ $y = -0.5x$

Algebra 1      Standardized Test Practice 5.4

NAME _____ CLASS _____ DATE _____

# Standardized Test Practice
## 5.5 The Standard and Point-Slope Forms

**TEST TAKING STRATEGY** Read the question carefully, know exactly what the question is asking.

1. **Multiple Choice** Which equation is written in point-slope form?
   - Ⓐ $y = 6x - 28$
   - Ⓑ $y - 2 = 6(x - 5)$
   - Ⓒ $6x - y = 28$
   - Ⓓ $y + 28 = \dfrac{2 + 28}{5 - 0}x$

2. **Multiple Choice** What is an equation in standard form of a line with slope $-2.5$ that contains point $(-4, 5)$?
   - Ⓐ $5x - 2y = 10$
   - Ⓑ $y = -2.5x - 5$
   - Ⓒ $5x + 2y = 10$
   - Ⓓ $5x + 2y = -10$

3. **Multiple Choice** The $x$- and $y$-intercepts of the graph of $7x - 3y = 42$ are:
   - Ⓐ 3 and 7
   - Ⓑ 14 and 6
   - Ⓒ $-14$ and 6
   - Ⓓ 7 and $-3$

4. **Multiple Choice** What are the values of $A$, $B$, and $C$ of the equation $3y + 8 = 2x - 4$ in standard form?
   - Ⓐ 2, 3, 12
   - Ⓑ $-2, 3, -12$
   - Ⓒ $-2, 3, 12$
   - Ⓓ $-2, -3, -12$

5. **Multiple Choice** Which equation *cannot* be written in slope-intercept form?
   - Ⓐ $x = -3$
   - Ⓑ $y = 2$
   - Ⓒ $x = 1.5y$
   - Ⓓ $x + y = -4$

6. **Multiple Choice** What is an equation in point-slope form of the line containing $(6, -1)$ and $(0, 5)$?
   - Ⓐ $y - 1 = 2 - x - 6$
   - Ⓑ $y - 5 = -(x - 0)$
   - Ⓒ $y + 1 = -x$
   - Ⓓ $y - 6 = 5(x - 0)$

**Quantitative Comparison** In Exercises 7–9, choose the letter of the statement below that is true about the quantities in Columns I and II.

A The number in Column I is greater.
B The number in Column II is greater.
C The two numbers are equal.
D The relationship cannot be determined from the given information.

| | Column I | Column II |
|---|---|---|
| 7. | the $x$-intercept of $y = 5x - 15$ | the slope of $6x - y = 11$ |
| | Ⓐ  Ⓑ | Ⓒ  Ⓓ |
| 8. | the $y$-intercept of $4x + 2y = 8$ | the slope of $4x - y = 7$ |
| | Ⓐ  Ⓑ | Ⓒ  Ⓓ |
| 9. | the slope of a horizontal line | the slope of $2x + 4y = -3$ |
| | Ⓐ  Ⓑ | Ⓒ  Ⓓ |

10. **Multiple Choice** Which line has the same slope as $2x - 3y = 7$?
    - Ⓐ $5x + y = 8$
    - Ⓑ $4x - 5y = 1$
    - Ⓒ $2x + 3y = 7$
    - Ⓓ $3y = 2x + 7$

11. **Multiple Choice** What is the standard form of the equation $y = 2$?
    - Ⓐ $y - 2 = 0$
    - Ⓑ $x + y = 2$
    - Ⓒ $0x + 1y = 2$
    - Ⓓ $y - 2 = x$

12. **Multiple Choice** What are the slope and $y$-intercept of the line $7x - 3y = 4$?
    - Ⓐ $\dfrac{7}{3}$ and $-\dfrac{4}{3}$
    - Ⓑ $-\dfrac{7}{3}$ and $\dfrac{4}{3}$
    - Ⓒ $-\dfrac{7}{3}$ and $-\dfrac{4}{3}$
    - Ⓓ $\dfrac{7}{3}$ and $\dfrac{4}{3}$

NAME _____ CLASS _____ DATE _____

# Standardized Test Practice
## 5.6 Parallel and Perpendicular Lines

**TEST TAKING STRATEGY**   Use context to help define an unknown word.

1. **Multiple Choice**  Which line is parallel to the line $6x - 3y = 7$?
   - Ⓐ $-2x - 2y = 7$
   - Ⓑ $x - 2y = 11$
   - Ⓒ $-2x + y = 8$
   - Ⓓ $4x + 2y - 0$

2. **Multiple Choice**  A line perpendicular to the line $6x - 3y = 7$ can be written as:
   - Ⓐ $4x + y = 2$
   - Ⓑ $x + 2y = 8$
   - Ⓒ $x - 2y = 5$
   - Ⓓ $2x - y = 14$

3. **Multiple Choice**  Which line containing the point $(0, -2)$ is parallel to $y = 1.5x + 6$?
   - Ⓐ $2y = 3x - 4$
   - Ⓑ $7x + 3y = -6$
   - Ⓒ $y = -2$
   - Ⓓ $2x + 3y = -4$

Use the graph for Exercises 4–6.

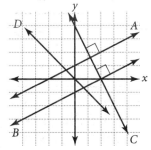

4. **Multiple Choice**  The slope of line $A$ is 0.5. What is the slope of line $C$?
   - Ⓐ $-0.5$
   - Ⓑ $0.5$
   - Ⓒ $-2$
   - Ⓓ $2$

5. **Multiple Choice**  The equation of line $D$ is $y = -1.5x$. What line goes through the origin and runs perpendicular to line $D$?
   - Ⓐ $y = \dfrac{3}{2}x$
   - Ⓑ $-2y = 3x$
   - Ⓒ $y = -\dfrac{2}{3}x$
   - Ⓓ $y = \dfrac{2}{3}x$

6. **Multiple Choice**  Which equation describes the line through the origin, parallel to lines $A$ and $B$?
   - Ⓐ $y = -0.5x$
   - Ⓑ $y = 0.5x$
   - Ⓒ $y = 2x$
   - Ⓓ $y = -2x$

**Quantitative Comparison**  In Exercises 7–8, choose the letter of the statement below that is true about the quantities in Columns I and II.

A  The number in Column I is greater.
B  The number in Column II is greater.
C  The two numbers are equal.
D  The relationship cannot be determined from the given information.

| | Column I | Column II |
|---|---|---|
| 7. | the slope of the line $5x - 2y = 8$ | the slope of a line perpendicular to $5x - 2y = 8$ |
| | Ⓐ   Ⓑ | Ⓒ   Ⓓ |
| 8. | the slope of a horizontal line | the slope of a line containing $(4, -2)$ |
| | Ⓐ   Ⓑ | Ⓒ   Ⓓ |

9. **Multiple Choice**  Side $AB$ of rectangle $ABCD$ is contained in the line shown. Which line could contain the side opposite in rectangle $ABCD$?

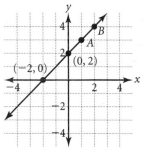

   - Ⓐ $y = -x$
   - Ⓑ $y = x - 2$
   - Ⓒ $y = 0.5x + 2$
   - Ⓓ $y = -0.5x + 2$

10. **Multiple Choice**  Which line could contain side $AD$?
    - Ⓐ $y = x + 6$
    - Ⓑ $y = -0.5x + 3$
    - Ⓒ $y = -x + 6$
    - Ⓓ $y = 0.5x$

# SAT/ACT Chapter Test
## Chapter 5  Linear Functions

**TEST TAKING STRATEGY**  Treat each answer like a true false question.

1. **Multiple Choice** Which point lies on the graph of $3x - y = 7$?
   - Ⓐ (2, 1)
   - Ⓑ (1, −4)
   - Ⓒ (−1, 10)
   - Ⓓ (4, −5)

2. **Multiple Choice** Which relation is a function?
   - Ⓐ {(6, 4), (5, 3), (6, 8), (4, 3)}
   - Ⓑ {(0, −3), (−2, −1), (0, −2), (1, −1)}
   - Ⓒ {(2, 6), (3, 6), (4, 2), (5, 2)}
   - Ⓓ {(11, 5), (12, 3), (11, 4), (16, 5)}

3. **Multiple Choice** What is the slope of the line that contains the points $Q(3, -1)$ and $P(-1, 7)$?
   - Ⓐ $\frac{1}{4}$
   - Ⓑ 2
   - Ⓒ $-\frac{1}{2}$
   - Ⓓ −2

4. **Multiple Choice** A line with slope of 1 contains the points $A(5, 2)$ and $B(-1, y)$. What is the value of $y$?
   - Ⓐ −6
   - Ⓑ −4
   - Ⓒ 8
   - Ⓓ 1

5. **Multiple Choice** If $y$ varies directly as $x$ and $y = 5$ when $x = 2$, what is the constant of variation?
   - Ⓐ 0.4
   - Ⓑ 10
   - Ⓒ 2.5
   - Ⓓ 7

6. **Multiple Choice** What is true for every direct-variation equation?
   - Ⓐ The graphs all have positive slope.
   - Ⓑ The graphs all have negative slope.
   - Ⓒ Each graph contains the origin.
   - Ⓓ The graphs are all vertical lines.

*Quantitative Comparison* In Exercises 7–9, choose the letter of the statement below that is true about the quantities in Columns I and II.

**A** The number in Column I is greater.
**B** The number in Column II is greater.
**C** The two numbers are equal.
**D** The relationship cannot be determined from the given information.

| Column I | Column II |
|---|---|
| 7. the $x$-intercept of line $y = 2x + 4$ | the $y$-intercept of line $y = 2x + 4$ |
| Ⓐ  Ⓑ | Ⓒ  Ⓓ |
| 8. $-\frac{2}{5}$ | the slope of the line $5y = -2x$ |
| Ⓐ  Ⓑ | Ⓒ  Ⓓ |
| 9. the slope of a line parallel to the line $5x - y = 5$ | the slope of a line perpendicular to the line $5x - y = 5$ |
| Ⓐ  Ⓑ | Ⓒ  Ⓓ |

10. **Multiple Choice** Which shows the standard form of the equation for the line containing points (3, 2) and (4, 0)?
    - Ⓐ $-2x - y = 8$
    - Ⓑ $-2x + y = 8$
    - Ⓒ $2x - y = -8$
    - Ⓓ $2x + y = 8$

11. **Multiple Choice** If a line crosses the axes at $x = 3$ and $y = 5$, what is the equation, in point-slope form?
    - Ⓐ $y + 5 = -\frac{5}{3}(x - 0)$
    - Ⓑ $y - 5 = \frac{-5}{3}(x - 0)$
    - Ⓒ $y - 5 = 0.6(x - 0)$
    - Ⓓ $y - 5 = -0.6(x - 0)$

NAME _____ CLASS _____ DATE _____

# Standardized Test Practice
## 6.1 Solving Inequalities

**TEST TAKING STRATEGY**  Look over the test before starting, use the test as an information tool.

1. **Multiple Choice** The sum of $x$ and 2.8 is greater than or equal to 15.4. Which inequality represents this situation?
   - Ⓐ $x \geq 15.4$
   - Ⓑ $x + 2.8 \geq 15.4$
   - Ⓒ $x + 2.8 > 15.4$
   - Ⓓ $x + 2.8 \leq 15.4$

2. **Multiple Choice** Which inequality describes the set of points graphed below?

   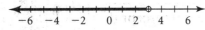

   - Ⓐ $x \leq 3$
   - Ⓑ $x > 3$
   - Ⓒ $x \geq 3$
   - Ⓓ $x < 3$

3. **Multiple Choice** What is the value of $a$ in the inequality $a + 12 \leq 16$.
   - Ⓐ $a \leq 4$
   - Ⓑ $a \leq 28$
   - Ⓒ $a \leq -4$
   - Ⓓ $a \geq 4$

4. **Multiple Choice** Which inequality has $n < -4$ as its solution?
   - Ⓐ $n - 2.6 < -1.4$
   - Ⓑ $n + 2.6 < -1.4$
   - Ⓒ $n + 3 < 1$
   - Ⓓ $n - 2 < -2$

5. **Multiple Choice** Which graph represents the solution of $y - 8 \leq 0$?

   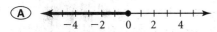

**Quantitative Comparison**  In Exercises 6–8, choose the letter of the statement below that is true about the quantities in Columns I and II.

A  The number in Column I is greater.
B  The number in Column II is greater.
C  The two numbers are equal.
D  The relationship cannot be determined from the given information.

| | Column I | Column II |
|---|---|---|
| 6. | the smallest solution of $x - 2 \geq 7$ | the greatest solution of $x + 3 \leq 12$ |
| | Ⓐ　　Ⓑ | Ⓒ　　Ⓓ |
| 7. | the greatest solution of $s + 1.5 \leq 6$ | the smallest solution of $s - 3 \geq 1.6$ |
| | Ⓐ　　Ⓑ | Ⓒ　　Ⓓ |
| 8. | a solution of $h + 3.2 < 8$ | a solution of $h - 5 > -12$ |
| | Ⓐ　　Ⓑ | Ⓒ　　Ⓓ |

9. **Multiple Choice** The temperature stayed below 0° all day. Which inequality represents this situation?
   - Ⓐ $t \geq 0$
   - Ⓑ $t < 0$
   - Ⓒ $t \leq 0$
   - Ⓓ $t > 0$

10. **Multiple Choice** Even with a 75¢ raise, Henry's salary is still less than $10 per hour. Which inequality best represents this situation?
    - Ⓐ $s + 0.75 < 10$
    - Ⓑ $s + 0.75 \leq 10$
    - Ⓒ $s - 0.75 < 10$
    - Ⓓ $s - 0.75 > 10$

NAME _____ CLASS _____ DATE _____

# Standardized Test Practice
## 6.2 Multistep Inequalities

**TEST TAKING STRATEGY** Look for math ideas that are presented in words.

1. **Multiple Choice** If $-3x \geq 4$, which of the following is true?

   Ⓐ $x \geq 12$   Ⓑ $x \geq -\dfrac{4}{3}$

   Ⓒ $x \leq -\dfrac{4}{3}$   Ⓓ $x \geq -12$

2. **Multiple Choice** Which best describes how to solve $-6 \geq \dfrac{t}{8}$?

   Ⓐ Divide both sides by 8 and reverse the sign.
   Ⓑ Multiply both sides by 8 and reverse the sign.
   Ⓒ Divide both sides by 8.
   Ⓓ Multiply both sides by 8.

3. **Multiple Choice** Solve $-3x + 7 < x - 2$ for $x$.

   Ⓐ $x < \dfrac{9}{4}$   Ⓑ $x > \dfrac{9}{4}$

   Ⓒ $x > \dfrac{9}{2}$   Ⓓ $x < -\dfrac{9}{2}$

4. **Multiple Choice** The graph shown below is the solution of which inequality?

   ←—+—+—+—+—+—●—+—+—+→
   $-6\ -4\ -2\ \ 0\ \ 2\ \ 4\ \ 6$

   Ⓐ $5 - 3x \leq -1$
   Ⓑ $3x \geq -6$
   Ⓒ $3x + 2 \geq 0$
   Ⓓ $3x - 6 \leq 0$

5. **Multiple Choice** Which inequality makes the statement true?

   $$\dfrac{n}{-4} + 2 > 5$$

   Ⓐ $n > -12$   Ⓑ $n > -28$

   Ⓒ $n < -12$   Ⓓ $n < -\dfrac{3}{4}$

**Quantitative Comparison** In Exercises 6–7, choose the letter of the statement below that is true about the quantities in Columns I and II.

A The number in Column I is greater.
B The number in Column II is greater.
C The two numbers are equal.
D The relationship cannot be determined from the given information.

| | Column I | Column II |
|---|---|---|
| 6. | the smallest solution of $\dfrac{a}{2} - 1 \geq 0$ | the greatest solution of $-3n \geq -15$ |
| | Ⓐ    Ⓑ | Ⓒ    Ⓓ |
| 7. | the solution of $-4x + 6 = -6$ | the greatest solution of $\dfrac{x}{-3} + 5 \geq 4$ |
| | Ⓐ    Ⓑ | Ⓒ    Ⓓ |

8. **Multiple Choice** The expenses for a fund-raising car wash are $75. At $4 per car wash, how many cars must be washed for the event to raise at least $100?

   Ⓐ 44 cars   Ⓑ 45 cars
   Ⓒ 43 cars   Ⓓ 7 cars

9. **Multiple Choice** The length, $l$, of a rectangle exceeds its width, $w$, by at least 2.5 centimeters. Which expresses this relationship?

   Ⓐ $l + 2.5 \geq w$
   Ⓑ $l - w \leq 2.5$
   Ⓒ $l \geq w + 2.5$
   Ⓓ $l + w \geq 2.5$

NAME _____ CLASS _____ DATE _____

# Standardized Test Practice
## 6.3 Compound Inequalities

**TEST TAKING STRATEGY**   Make sure you know exactly what the question is asking.

1. **Multiple Choice** The graph shown below is the graph of which inequality?

   Ⓐ $p > -1$  AND  $p \leq 1$
   Ⓑ $p \geq -1$  AND  $p < 1$
   Ⓒ $p > -1$  OR  $p \leq 1$
   Ⓓ $p < -1$  OR  $p \leq 1$

2. **Multiple Choice** Which is the graph of $-2 \leq y + 1 \leq 7$?

   Ⓐ
   Ⓑ
   Ⓒ
   Ⓓ

3. **Multiple Choice** Which is the solution of $0 \leq 2x + 4 \leq 8$?

   Ⓐ $x \leq 2$        Ⓑ $-2 \leq x \leq 2$
   Ⓒ $x \leq -2$       Ⓓ $-2 \leq x \leq 6$

4. **Multiple Choice** Which number is *not* a solution of $7y - 3 \leq 11$ OR $2y + 4 \geq 10$?

   Ⓐ 2        Ⓑ 2.5
   Ⓒ 3        Ⓓ 3.5

5. **Multiple Choice** Which inequality has the entire number line as the graph of its solution set?

   Ⓐ $2z - 1 \geq 0$ AND $z + 4 < -3$
   Ⓑ $3z - 2 \geq -2$ AND $-2z + 4 > 0$
   Ⓒ $3z - 2 \geq -2$ OR $-2z + 4 > 0$
   Ⓓ $2z - 1 \geq 0$ OR $z + 4 < -3$

**Quantitative Comparison** In Exercises 6–8, choose the letter of the statement below that is true about the quantities in Columns I and II.

**A** The number in Column I is greater.
**B** The number in Column II is greater.
**C** The two numbers are equal.
**D** The relationship cannot be determined from the given information.

| | Column I | Column II |
|---|---|---|
| 6. | the greatest solution of $-4 \leq x \leq 5$ | the greatest solution of $0 \leq x \leq 4$ |
| | Ⓐ   Ⓑ | Ⓒ   Ⓓ |
| 7. | the solution of $-5x + 3 \geq 3$ AND $3x - 7 \geq -7$ | the solution of $6x - 4 \leq 8$ AND $x \geq 2$ |
| | Ⓐ   Ⓑ | Ⓒ   Ⓓ |
| 8. | the least solution of $-5 \leq x \leq 0$ | the least solution of $-5 < x < 0$ |
| | Ⓐ   Ⓑ | Ⓒ   Ⓓ |

9. **Multiple Choice** At a fair, children under 12 and adults over 60, $x$, pay half-price for admission. Which inequality describes this age category?

   Ⓐ $x < 12$  AND  $x > 60$
   Ⓑ $x > 12$  AND  $x < 60$
   Ⓒ $12 < x < 60$
   Ⓓ $x < 12$  OR  $x > 60$

10. **Multiple Choice** Water changes to ice below 0°C and to steam above 100°C. When is water liquid?

    Ⓐ $t \leq 0$ OR $t \geq 100$
    Ⓑ $t \geq 0$ OR $t \geq 100$
    Ⓒ $0 \leq t \leq 100$
    Ⓓ $t \leq 0$ AND $t \leq 100$

NAME _____ CLASS _____ DATE _____

# Standardized Test Practice
## 6.4 Absolute-Value Functions

**TEST TAKING STRATEGY**  Use number sense to eliminate unreasonable choices.

1. **Multiple Choice**  Evaluate $|-10 + 8|$.
   - Ⓐ $-2$
   - Ⓑ $18$
   - Ⓒ $2$
   - Ⓓ $-18$

2. **Multiple Choice**  What is the range of $y = |x| - 3$?
   - Ⓐ $y \geq -3$
   - Ⓑ $y > 0$
   - Ⓒ $y \geq 0$
   - Ⓓ $y \geq -3$

3. **Multiple Choice**  Which function is shown in the graph?

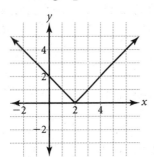

   - Ⓐ $y = |x| + 2$
   - Ⓑ $y = |x - 2|$
   - Ⓒ $y = |x| - 2$
   - Ⓓ $y = |x + 2|$

4. **Multiple Choice**  What kind of transformation of the graph of $y = |x|$ is the graph of $y = |x| + 5$?
   - Ⓐ a vertical translation downward
   - Ⓑ a reflection
   - Ⓒ a horizontal translation to the left
   - Ⓓ a vertical translation upward

**Quantitative Comparison**  In Exercises 5–8, choose the letter of the statement below that is true about the quantities in Columns I and II.

A  The number in Column I is greater.
B  The number in Column II is greater.
C  The two numbers are equal.
D  The relationship cannot be determined from the given information.

| | Column I | Column II |
|---|---|---|
| 5. | $\|0 - 5\|$ | $\|0 - 10\|$ |
| | Ⓐ  Ⓑ | Ⓒ  Ⓓ |
| 6. | $\|x + 6\|$ if $x = -6$ | $\|x + 6\|$ if $x = 6$ |
| | Ⓐ  Ⓑ | Ⓒ  Ⓓ |
| 7. | $\frac{1}{2}\|x\| + 3$ if $x = -4$ | $\|x\|$ if $x = -4$ |
| | Ⓐ  Ⓑ | Ⓒ  Ⓓ |
| 8. | $3\|x - 2\|$ if $x = -5$ | $5\|x - 1\|$ if $x = -3$ |
| | Ⓐ  Ⓑ | Ⓒ  Ⓓ |

9. **Multiple Choice**  Which point is not on the graph of $y = -|x|$?
   - Ⓐ $(0, 0)$
   - Ⓑ $(3, 3)$
   - Ⓒ $(-2, -2)$
   - Ⓓ $(3, -3)$

10. **Multiple Choice**  Which function describes the graph of $y = |x|$ translated 3 units to the left and 1 unit down?
    - Ⓐ $y = |x - 3| - 1$
    - Ⓑ $y = |x + 1| - 3$
    - Ⓒ $y = -|x + 3| - 1$
    - Ⓓ $y = |x + 3| - 1$

NAME _____ CLASS _____ DATE _____

# Standardized Test Practice
## 6.5 Absolute-Value Equations and Inequalities

**TEST TAKING STRATEGY** Look at all of the answer choices before choosing one.

**1. Multiple Choice** Solve $|5x - 1| = 11$.

- Ⓐ 2.4
- Ⓑ −2.4 and 2
- Ⓒ 2.4 and −2
- Ⓓ 2.4 and 2

**2. Multiple Choice** Solve $|3y - 5| \le 2$.

- Ⓐ $y < 1$ OR $y \ge 2\frac{1}{3}$
- Ⓑ $1 \le y \le 2\frac{1}{3}$
- Ⓒ $-1 \le y \le 2\frac{1}{3}$
- Ⓓ $y > -1$ OR $y < 2\frac{1}{3}$

**3. Multiple Choice** The graph shown is the solution interval for which inequality?

- Ⓐ $|a - 3| \le 4$
- Ⓑ $|a - 3| \le 7$
- Ⓒ $|a + 3| \le 4$
- Ⓓ $|a - 3| \le -1$

**4. Multiple Choice** Solve $|2t + 3| > 1$.

- Ⓐ $t < -1$ OR $t > 2$
- Ⓑ $t > -1$ AND $t < -2$
- Ⓒ $t > 1$ AND $t < -1$
- Ⓓ $t > -1$ OR $t < -2$

**5. Multiple Choice** Which best describes the graph of the solution set of $|t + 0.02| \le 0.001$?

- Ⓐ all numbers between 0.019 and 0.021
- Ⓑ $-0.021 \le t \le 0.019$
- Ⓒ all numbers between 0.01 and 0.02
- Ⓓ $-0.021 < t < 0.019$

**Quantitative Comparison** In Exercises 6–8, choose the letter of the statement below that is true about the quantities in Columns I and II.

- **A** The number in Column I is greater.
- **B** The number in Column II is greater.
- **C** The two numbers are equal.
- **D** The relationship cannot be determined from the given information.

| | Column I | Column II |
|---|---|---|
| 6. | the upper boundary of $\|x - 4\| \le 6$ | the upper boundary of $\|2x - 3\| \le 1$ |
| | Ⓐ    Ⓑ | Ⓒ    Ⓓ |
| 7. | the lower boundary of $\|x - 4\| \le 6$ | the lower boundary of $\|2x - 3\| \le 1$ |
| | Ⓐ    Ⓑ | Ⓒ    Ⓓ |
| 8. | the upper boundary of $\|2x - 3\| \ge 0$ | the lower boundary of $\|2x - 3\| \ge 0$ |
| | Ⓐ    Ⓑ | Ⓒ    Ⓓ |

**9. Multiple Choice** The margin of error in a political survey is ±2.5%. The survey indicates that candidate A is preferred by 45% of the voters. What percent of the voters might actually support candidate A?

- Ⓐ between 47% and 55%
- Ⓑ about 47.5%
- Ⓒ between 42.5% and 47.5%
- Ⓓ about 42.5%

Algebra 1        Standardized Test Practice 6.5

# SAT/ACT Chapter Test

## Chapter 6  Inequalities and Absolute Value

**TEST TAKING STRATEGY**  Even when you find your answer choice, be sure to check your work.

1. **Multiple Choice** Which solution matches the inequality given below?

   $$d - 2.4 \leq 8.2$$

   Ⓐ $d \leq 5.8$  Ⓑ $d \leq 10.6$
   Ⓒ $d \leq -5.8$  Ⓓ $d \leq 6.6$

2. **Multiple Choice** Which inequality has the solution shown in the graph?

   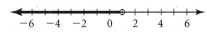

   Ⓐ $y - 2 < 3$
   Ⓑ $y + 2 \leq 3$
   Ⓒ $y + 2 < 3$
   Ⓓ $y - 2 \leq 3$

3. **Multiple Choice** Which is the graph of the solution of $6 - 2b \leq b + 3$?

   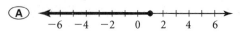

4. **Multiple Choice** Which best describes the solution to $-8 \leq 2x - 6 \leq 4$?

   Ⓐ $-7 \leq x \leq 5$  Ⓑ $x \geq -1$
   Ⓒ $x \leq 5$  Ⓓ $-1 \leq x \leq 5$

5. **Multiple Choice** Which inequality has the solution $x < -0.5$ OR $x \geq 0.5$?

   Ⓐ $4x + 2 < 0$ OR $3x + 1.5 \geq 3$
   Ⓑ $x + 1 \geq 1$ OR $2x + 1 > 1$
   Ⓒ $4x + 2 < 0$ OR $3x - 1.5 \geq 3$
   Ⓓ $2x \geq 2$ OR $x - 1 \geq -0.5$

*Quantitative Comparison*  In Exercises 6–9, choose the letter of the statement below that is true about the quantities in Columns I and II.

**A** The number in Column I is greater.
**B** The number in Column II is greater.
**C** The two numbers are equal.
**D** The relationship cannot be determined from the given information.

| | Column I | Column II |
|---|---|---|
| 6. | $|3 - 12|$ | $|12 - 3|$ |
| | Ⓐ  Ⓑ | Ⓒ  Ⓓ |
| 7. | the upper boundary of $|x + 7.5| \leq 8$ | the lower boundary of $|x - 12| \leq 1$ |
| | Ⓐ  Ⓑ | Ⓒ  Ⓓ |
| 8. | $|x + 3|$ when $x = -4$ | $|x + 3|$ when $x = 0$ |
| | Ⓐ  Ⓑ | Ⓒ  Ⓓ |
| 9. | the distance between $-4$ and 5 on a number line | the distance between $-4$ and $-12$ on a number line |
| | Ⓐ  Ⓑ | Ⓒ  Ⓓ |

10. **Multiple Choice** One factor of a number written in scientific notation must be a solution of $1 \leq x < 10$. Which number is *not* a solution of this inequality?

    Ⓐ 1.4  Ⓑ 8.7
    Ⓒ 9.9  Ⓓ 0.5

11. **Multiple Choic** Which inequality is the solution of $x + 5 \geq 5x - 7$?

    Ⓐ $x \leq 3$  Ⓑ $x \geq 3$
    Ⓒ $x \geq -3$  Ⓓ $x \leq -2$

NAME _____ CLASS _____ DATE _____

# Standardized Test Practice
## 7.1 Graphing Systems of Equations

**TEST TAKING STRATEGY**  Read each problem carefully.

1. **Multiple Choice**  Which value is the $x$-coordinate of the intersection of $3x + y = 15$ and $x + 2y = 10$?
   - (A) 3
   - (B) 4
   - (C) $-3$
   - (D) $-4$

2. **Multiple Choice**  In which quadrant does the graph of the following system intersect?
   $$\begin{cases} y = 3x + 1 \\ 2x + y = -9 \end{cases}$$
   - (A) I
   - (B) II
   - (C) III
   - (D) IV

3. **Multiple Choice**  Which coordinate point is the solution to the system?
   $$\begin{cases} 3x + 2y = 8 \\ x - y = 1 \end{cases}$$
   - (A) $(2, 1)$
   - (B) $(2, 0.5)$
   - (C) $(-2, -1)$
   - (D) $(3, 2)$

4. **Multiple Choice**  Which of the following *cannot* be the solution to a system of linear equations?
   - (A) $(0, 0)$
   - (B) $(0.73, 0.56)$
   - (C) $(3\frac{1}{9}, 2)$
   - (D) $(2, 9)$ and $(5, 1)$

5. **Multiple Choice**  What is the $y$-coordinate of the intersection of $2x + y = 9$ and $x + 2y = 15$?
   - (A) 4.2
   - (B) 1
   - (C) 8
   - (D) 7

6. **Multiple Choice**  Find the coordinates of the solution to the system.
   $$\begin{cases} 2x + y = 2 \\ y - 3x = 9 \end{cases}$$
   - (A) $(1.5, 4.5)$
   - (B) $(-7.0, -12.0)$
   - (C) $(-2.65, 1.04)$
   - (D) $(4.8, -1.4)$

**Quantitative Comparison**  In Exercises 7 and 8, choose the letter of the statement below that is true about the quantities in Columns I and II.

A  The number in Column I is greater.
B  The number in Column II is greater.
C  The two numbers are equal.
D  The relationship cannot be determined from the given information.

| | Column I | Column II |
|---|---|---|
| 7. | $\begin{cases} 2x - y = 3 \\ x + y = -9 \end{cases}$ | |
| | $x$ | $y$ |
| | (A)   (B) | (C)   (D) |
| 8. | Jane has $7 in dimes and quarters. There are 40 coins in all. | |
| | number of nickels | number of quarters |
| | (A)   (B) | (C)   (D) |

9. **Multiple Choice**  Where is the intersection of this system located?
   $$\begin{cases} x + y = 5 \\ x - y = -5 \end{cases}$$
   - (A) in a quadrant
   - (B) on the $x$-axis
   - (C) on the $y$-axis
   - (D) the origin

10. **Multiple Choice**  Which ordered pair is a solution to the system graphed?
    - (A) $(-1, 2)$
    - (B) $(-2, 1)$
    - (C) $(1, -2)$
    - (D) $(1, -2)$

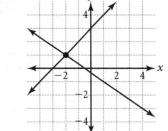

Algebra 1                                                     Standardized Test Practice 7.1    43

NAME _____ CLASS _____ DATE _____

# Standardized Test Practice
## 7.2 The Substitution Method

**TEST TAKING STRATEGY** Make an inference to fill in missing information.

1. **Multiple Choice** Solve the system using substitution.
$$\begin{cases} x + 2y = 11 \\ 3x + 2y = 13 \end{cases}$$
   (A) $(1, 5)$  (B) $(16, -2.5)$
   (C) $(-1, 6)$  (D) $(12, -0.5)$

2. **Multiple Choice** To solve the system by the substitution method, which would *not* be an appropriate first step?
$$\begin{cases} x + 5y = -25 \\ 3x + y = 9 \end{cases}$$
   (A) $x = -25 - 5y$  (B) $y = -5 - x$
   (C) $y = 9 - 3x$  (D) $y = -0.5x - 5$

3. **Multiple Choice** What is the value of $x$ in the solution of the system?
$$\begin{cases} 8x + 3y = 19 \\ 2x + y = 6 \end{cases}$$
   (A) 6  (B) 2
   (C) $\frac{1}{6}$  (D) $\frac{1}{2}$

4. **Multiple Choice** What is the resulting equation when the second equation in the system is substituted into the first equation?
$$\begin{cases} 6x - 4y = 11 \\ y = 3x - 2 \end{cases}$$
   (A) $6x + 8 = 11$  (B) $-6x - 2 = 11$
   (C) $-6x - 8 = 11$  (D) $-6x + 8 = 11$

5. **Multiple Choice** Using substitution, a solution to a system of linear equations:
   (A) is correct if it satisfies either of the equations.
   (B) is the exact intersection of the lines.
   (C) is a reasonable estimate.
   (D) cannot be checked without a graph.

**Quantitative Comparison** In Exercises 6–8, choose the letter of the statement below that is true about the quantities in Columns I and II.

**A** The number in Column I is greater.
**B** The number in Column II is greater.
**C** The two numbers are equal.
**D** The relationship cannot be determined from the given information.

| | Column I | Column II |
|---|---|---|
| 6. | $\begin{cases} 3x - 5y = 8 \\ x + 2y = -1 \end{cases}$ | |
| | $x$ | $y$ |
| | (A) (B) | (C) (D) |

7. The measure of angle $A$ is 39° less than twice the measure of angle $B$.
   | $m\angle A$ | $m\angle B$ |
   |---|---|
   | (A) (B) | (C) (D) |

8. | $x$-coordinate of the intersection of $\begin{cases} y = 6 \\ x + 2y = 16 \end{cases}$ | $y$ coordinate of the intersection of $\begin{cases} 2x + y = 16 \\ x = 6 \end{cases}$ |
   |---|---|
   | (A) (B) | (C) (D) |

9. **Multiple Choice** To use substitution, you first select an equation and solve it for a variable. What is the logical choice for this system?
$$\begin{cases} 5x + 3y = 45 \\ x - 6y = 40 \end{cases}$$
   (A) first equation for $x$
   (B) first equation for $y$
   (C) second equation for $x$
   (D) second equation for $y$

# Standardized Test Practice
## 7.3 The Elimination Method

**TEST TAKING STRATEGY** Use number sense to eliminate unreasonable choices.

1. **Multiple Choice** In order to solve the system by elimination, what quantity should be used to multiply each side of the first equation before applying the Addition Property of Equality?
$$\begin{cases} 2x + y = 5 \\ 3x + 2y = 10 \end{cases}$$
   - (A) $-3$
   - (B) $-2$
   - (C) $2$
   - (D) $3$

2. **Multiple Choice** Solve the system by elimination.
$$\begin{cases} 5x + 2y = 7 \\ 3x - 4y = -1 \end{cases}$$
   - (A) $(-1, 1)$
   - (B) $(1, -1)$
   - (C) $(1, 1)$
   - (D) $(-1, 6)$

3. **Multiple Choice** In the system, what smallest possible factor of 48 can be used to multiply the first equation to eliminate $x$?
$$\begin{cases} 6x + 5y = 13 \\ 8x - 3y = -31 \end{cases}$$
   - (A) $2$
   - (B) $4$
   - (C) $5$
   - (D) $8$

4. **Multiple Choice** There exist two numbers such that the first plus twice the second equals 1, and 3 times the first plus the second is 8. What is the sum of these two numbers?
   - (A) $2$
   - (B) $3$
   - (C) $5$
   - (D) $8$

5. **Multiple Choice** What is the value of $x$ in the system?
$$\begin{cases} 8x + 3y = 2 \\ 11x + 2y = -10 \end{cases}$$
   - (A) $-6$
   - (B) $-2$
   - (C) $2$
   - (D) $6$

**Quantitative Comparison** In Exercises 6 and 7, choose the letter of the statement below that is true about the quantities in Columns I and II.

A The number in Column I is greater.
B The number in Column II is greater.
C The two numbers are equal.
D The relationship cannot be determined from the given information.

| | Column I | Column II |
|---|---|---|
| 6. | $\begin{cases} 5x + 2y = 56 \\ 10x - 3y = 56 \end{cases}$ | |
| | $x$ | $y$ |
| | (A)  (B) | (C)  (D) |
| 7. | The $x$-value in: $\begin{cases} y = 2x \\ x + y = 9 \end{cases}$ | The $x$-value in: $\begin{cases} 2x - y = 9 \\ 5x + 2y = 27 \end{cases}$ |
| | (A)  (B) | (C)  (D) |

8. **Multiple Choice** What is the value of $y$ in the system?
$$\begin{cases} 3x + 5y = 12 \\ 3x - 5y = -36 \end{cases}$$
   - (A) $-4$
   - (B) $-2.4$
   - (C) $0$
   - (D) $4.8$

9. **Multiple Choice** What method is the best choice to solve the sytem?
$$\begin{cases} x + 8y = 41 \\ x = 9 \end{cases}$$
   - (A) graphing
   - (B) elimination
   - (C) substitution
   - (D) guess and check

NAME _____ CLASS _____ DATE _____

# Standardized Test Practice
## 7.4 Consistent and Inconsistent Systems

**TEST TAKING STRATEGY**  Restate the question to verify that you answered it.

1. **Multiple Choice** For which of the following systems do the graphs of the equations have no points in common?

   Ⓐ $\begin{cases} x + y = 3 \\ -x + 2y = 5 \end{cases}$  Ⓑ $\begin{cases} x + y = 3 \\ -x - y = 0 \end{cases}$

   Ⓒ $\begin{cases} x + y = 3 \\ 3x + 3y = 5 \end{cases}$  Ⓓ $\begin{cases} x + y = 3 \\ x + y = -2 \end{cases}$

2. **Multiple Choice** A system of equations representing two different vertical lines is:

   Ⓐ consistent independent.
   Ⓑ consistent dependent.
   Ⓒ inconsistent.
   Ⓓ inconsistent independent.

3. **Multiple Choice** Which equation creates a dependent system with $y = 0.5x - 3$?

   Ⓐ $x - 2y = 6$   Ⓑ $y = 0.5x + 4$
   Ⓒ $y = 2x - 3$   Ⓓ $x + 2y = -6$

4. **Multiple Choice** Choose the statement that is always true if a system of equations is independent.

   Ⓐ The lines intersect the y-axis at the same point.
   Ⓑ The lines have different slopes.
   Ⓒ The lines pass through the origin.
   Ⓓ The lines intersect the x-axis at the same point.

5. **Multiple Choice** Classify:

   $\begin{cases} 5x + y = 19 \\ x + 5y = 23 \end{cases}$

   Ⓐ consistent, independent
   Ⓑ inconsistent
   Ⓒ inconsistent, dependent
   Ⓓ dependent

6. **Multiple Choice** Which system is dependent?

   Ⓐ $\begin{cases} x + y = 7 \\ x - y = 3 \end{cases}$  Ⓑ $\begin{cases} x + 2y = 6 \\ x = 2 - 2y \end{cases}$

   Ⓒ $\begin{cases} 3x - 2y = 6 \\ 4y = 6x - 12 \end{cases}$  Ⓓ $\begin{cases} x - 2y = -7 \\ 2x - 14 = 4y \end{cases}$

**Quantitative Comparison** In Exercises 7–9, choose the letter of the statement below that is true about the quantities in Columns I and II.

**A** The number in Column I is greater.
**B** The number in Column II is greater.
**C** The two numbers are equal.
**D** The relationship cannot be determined from the given information.

|   | Column I | Column II |
|---|----------|-----------|
| 7. | $\begin{cases} x + 2y = 28 \\ \frac{1}{2}x + \frac{1}{3}y = 8 \end{cases}$ when $x = 2$ | |
|   | first equation's y-value | second equation's y-value |
|   | Ⓐ  Ⓑ | Ⓒ  Ⓓ |
| 8. | number of solutions | number of solutions |
|   | $\begin{cases} 3x + 5y = 20 \\ -2x + 5y = 20 \end{cases}$ | $\begin{cases} y = -6x + 2 \\ 12x + 2y = 5 \end{cases}$ |
|   | Ⓐ  Ⓑ | Ⓒ  Ⓓ |
| 9. | number of solutions in an inconsistent system | number of solutions in a dependent system |
|   | Ⓐ  Ⓑ | Ⓒ  Ⓓ |

NAME _____ CLASS _____ DATE _____

# Standardized Test Practice
## 7.5 Systems of Inequalities

**TEST TAKING STRATEGY** Look for obvious distracters and eliminate them as choices.

1. *Multiple Choice* What is the boundary line equation of the graph of the inequality $2x - y < 7$?
   - (A) $y = -2x + 7$
   - (B) $y \geq 2x + 7$
   - (C) $y = 2x - 7$
   - (D) $y \geq 2x - 7$

2. *Multiple Choice* Which point belongs to the solution set of the system below?
   $$\begin{cases} y \leq 3x + 5 \\ x > -2 \end{cases}$$
   - (A) $(-3, -4)$
   - (B) $(-6, 0)$
   - (C) $(0, 6)$
   - (D) $(6, 0)$

3. *Multiple Choice* Which description best describes the graph of
   $2x - 7y \geq 14$?
   - (A) solid line and region below it
   - (B) solid line and region above it
   - (C) dashed line and region above it
   - (D) dashed line and region below it

4. *Multiple Choice* Identify the system whose graph is shown below.

   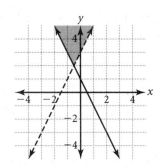

   - (A) $\begin{cases} y < 2x + 3 \\ y \geq -2x + 1 \end{cases}$
   - (B) $\begin{cases} y > 2x + 3 \\ y \geq -2x + 1 \end{cases}$
   - (C) $\begin{cases} y > 2x + 3 \\ y \leq -2x + 1 \end{cases}$
   - (D) $\begin{cases} y < 2x + 3 \\ y \leq -2x + 1 \end{cases}$

*Quantitative Comparison* In Exercises 5–7, choose the letter of the statement below that is true about the quantities in Columns I and II.

A The number in Column I is greater.
B The number in Column II is greater.
C The two numbers are equal.
D The relationship cannot be determined from the given information.

| | Column I | | Column II | |
|---|---|---|---|---|
| 5. | | $2x - y > -3$ | | |
| | | $y$ | $3 + 2x$ | |
| | (A) | (B) | (C) | (D) |
| 6. | the value of $-\frac{1}{2}x + \frac{1}{3}y$ given $(16, 9)$ | | $-5$ | |
| | (A) | (B) | (C) | (D) |
| 7. | number of solutions | | | |
| | $\begin{cases} y \leq 5 \\ y > x - 3 \end{cases}$ | | $\begin{cases} y < x - 5 \\ y \geq x + 2 \end{cases}$ | |
| | (A) | (B) | (C) | (D) |

8. *Multiple Choice* Which point satisfies only one of the inequalities in
   $$\begin{cases} x + y \geq 5 \\ y < 2x - 4 \end{cases}?$$
   - (A) $(3, 2)$
   - (B) $(9, 0)$
   - (C) $(7, 3)$
   - (D) $(8, -1)$

9. *Multiple Choice* The graph of $\begin{cases} x > 0 \\ y < 0 \end{cases}$ consist of all solutions in which quadrant?
   - (A) I
   - (B) II
   - (C) III
   - (D) IV

Algebra 1      Standardized Test Practice 7.5    47

# Standardized Test Practice
## 7.6 Classic Puzzles in Two Variables

**TEST TAKING STRATEGY** Look for math ideas that are presented in word form.

1. **Multiple Choice** Kay is $x$ years old and her father is $y$ years old. Seven years ago, he was 5 times as old as she was. Which equation is true of their age relationship 7 years ago?
   - Ⓐ $y = 5x$
   - Ⓑ $y - 7 = 5x$
   - Ⓒ $y = 5(x - 7)$
   - Ⓓ $y - 7 = 5(x - 7)$

2. **Multiple Choice** Bob is 5 years older than Ron. Twice Bob's age added to 4 times Ron's age is 64. How old is Bob?
   - Ⓐ 9
   - Ⓑ 14
   - Ⓒ 8
   - Ⓓ 7

3. **Multiple Choice** If a plane has an air speed of $x$ miles per hour and a tail wind of 50 miles per hour, how many hours will it take to fly 350 miles?
   - Ⓐ $\dfrac{350}{x + 50}$
   - Ⓑ $\dfrac{350}{x - 50}$
   - Ⓒ 7
   - Ⓓ $7x$

4. **Multiple Choice** Suppose you can row a boat 8 miles downstream in the same amount of time that you can row only 4 miles upstream. What is the speed of the current?
   - Ⓐ 1 mile per hour
   - Ⓑ 2 miles per hour
   - Ⓒ 4 miles per hour
   - Ⓓ 6 miles per hour

5. **Multiple Choice** Sue has nickels and quarters totaling $6.50. There are 4 more nickels than quarters. How many coins are there?
   - Ⓐ 21
   - Ⓑ 22
   - Ⓒ 25
   - Ⓓ 46

6. **Multiple Choice** How many ounces of pure (100%) alcohol solution should be mixed with a 10% alcohol solution to produce 9 ounces of a 40% solution?
   - Ⓐ 3
   - Ⓑ 4
   - Ⓒ 5
   - Ⓓ 6

*Quantitative Comparison* In Exercises 7–9, choose the letter of the statement below that is true about the quantities in Column I and II.

A The number in Column I is greater.
B The number in Column II is greater.
C The two numbers are equal.
D The relationship cannot be determined from the given information.

| Column I | Column II |
|---|---|

7. The sum of Jack's age and Jill's age is 58. Twice Jack's age added to 3 times Jill's age is 149.

   Jack's age | Jill's age

   Ⓐ　Ⓑ　Ⓒ　Ⓓ

8. The sum of the digits of a 2-digit number is 15. When 9 is added to the number, the digits are reversed.

   original tens digit | original ones digit

   Ⓐ　Ⓑ　Ⓒ　Ⓓ

9. Tasha has 46 coins valued at $3.45. She only has nickels and dimes.

   number of nickels | number of dimes

   Ⓐ　Ⓑ　Ⓒ　Ⓓ

NAME _____ CLASS _____ DATE _____

# SAT/ACT Chapter Test

### Chapter 7  Systems of Equations and Inequalities

**TEST TAKING STRATEGY**  Use context to help define an unknown word.

1. **Multiple Choice**  If the system is graphed, what is the solution?
$$\begin{cases} x + 3y = 2 \\ 2x - y = 7 \end{cases}$$
   - Ⓐ $(3.3, -0.4)$
   - Ⓑ $(3.8, -0.7)$
   - Ⓒ $(3.3, 0.4)$
   - Ⓓ $(3.8, 0.7)$

2. **Multiple Choice**  Using substitution to solve the system, which describes the exact solution?
$$\begin{cases} 7x + y = 5 \\ 14x - 7y = -8 \end{cases}$$
   - Ⓐ $(0.4, 2)$
   - Ⓑ $(2, 0.4)$
   - Ⓒ $(2, \frac{3}{7})$
   - Ⓓ $(\frac{3}{7}, 2)$

3. **Multiple Choice**  Which description best fits the graph of $3x + 2y > 5$?
   - Ⓐ solid line, slope $-\frac{2}{3}$; region with $(6, 1)$
   - Ⓑ solid line, slope $-\frac{3}{2}$; region with $(-1, 2)$
   - Ⓒ dashed line, slope $-\frac{2}{3}$; region with $(5, 0)$
   - Ⓓ dashed line, slope $-\frac{3}{2}$; region with $(1, 2)$

4. **Multiple Choice**  In what quadrants are the solutions to the system?
$$\begin{cases} 2y - x \geq 0 \\ x + y \leq 3 \end{cases}$$
   - Ⓐ I and II
   - Ⓑ I, II and III
   - Ⓒ I and IV
   - Ⓓ I, III and IV

5. **Multiple Choice**  Almonds cost $3.99 per pound and peanuts cost $1.79 per pound. If Jon makes 15 pounds of a nut mixture, worth $2.67 per pound, how many pounds of peanuts did he use?
   - Ⓐ 5 pounds
   - Ⓑ 6 pounds
   - Ⓒ 9 pounds
   - Ⓓ 10 pounds

6. **Multiple Choice**  The graph of the system below can be described as:
$$\begin{cases} 3x + 6y = 9 \\ 5x + 10y = 15 \end{cases}$$
   - Ⓐ parallel lines
   - Ⓑ the same line
   - Ⓒ vertical lines
   - Ⓓ a point

**Quantitative Comparison**  In Exercises 7–8, choose the letter of the statement below that is true about the quantities in Column I and II.

**A**  The number in Column I is greater.
**B**  The number in Column II is greater.
**C**  The two numbers are equal
**D**  The relationship cannot be determined from the given information.

| Column I | Column II |
|---|---|
| 7. $\begin{cases} 2x + 3y = -30 \\ 7x - 4y = -18 \end{cases}$ | |
| $x$ | $y$ |
| Ⓐ   Ⓑ | Ⓒ   Ⓓ |

8. The sum of Tom's age and Vanessa's age is 72. In 9 years the sum of their ages will be twice Tom's present age.

| Tom's age now | Vanessa's age now |
|---|---|
| Ⓐ   Ⓑ | Ⓒ   Ⓓ |

9. **Multiple Choice**  Use the elimination method to find the value of $x$ in the system.
$$\begin{cases} 3x + 4y = 11 \\ 7x + 5y = 4 \end{cases}$$
   - Ⓐ $-5$
   - Ⓑ $-3$
   - Ⓒ $3$
   - Ⓓ $5$

Algebra 1

NAME _____ CLASS _____ DATE _____

# Standardized Test Practice
## 8.1 Laws of Exponents: Multiplying Monomials

**TEST TAKING STRATEGY**  Use the process of elimination to disregard unreasonable answers.

1. **Multiple Choice**  What is 6 raised to the third power?
   - Ⓐ 18
   - Ⓑ 216
   - Ⓒ 3
   - Ⓓ 108

2. **Multiple Choice**  Which of the following equals 4096?
   - Ⓐ $4^4$
   - Ⓑ $6^4$
   - Ⓒ $8^4$
   - Ⓓ $12^4$

3. **Multiple Choice**  What is the value of $a^4$ when $a = 3$?
   - Ⓐ 12
   - Ⓑ 36
   - Ⓒ 72
   - Ⓓ 81

4. **Multiple Choice**  Which mathematical operation is performed on $x$ and $y$ to simplify the expression $2^x \cdot 2^y$?
   - Ⓐ addition
   - Ⓑ subtraction
   - Ⓒ division
   - Ⓓ multiplication

5. **Multiple Choice**  Which of the following is equivalent to $(6^4 \cdot 6^8)$?
   - Ⓐ $36^{32}$
   - Ⓑ $6^{12}$
   - Ⓒ $6^{32}$
   - Ⓓ 1152

6. **Multiple Choice**  Which expression does *not* equal 100,000?
   - Ⓐ $10^5$
   - Ⓑ $10^5 \times 10^0$
   - Ⓒ 100(1000)
   - Ⓓ $10^5 \times 10$

7. **Multiple Choice**  Which of the following is equivalent to $8a^4(5a^3)$?
   - Ⓐ $40a^7$
   - Ⓑ $13a^7$
   - Ⓒ $40a^{12}$
   - Ⓓ $40a$

8. **Multiple Choice**  After the expression is simplified, to what power is $b$ raised?
   $$(-4a^7b^6)(3a^2b^3c^2)$$
   - Ⓐ 18
   - Ⓑ −9
   - Ⓒ 6
   - Ⓓ 9

9. **Multiple Choice**  If $z = 2$, what is the coefficient of the expression $(2x^2y^3z)(4xy^2z^4)$ when simplified?
   - Ⓐ 16
   - Ⓑ 64
   - Ⓒ 32
   - Ⓓ 256

*Quantitative Comparison*  In Exercises 10–13, choose the letter of the statement below that is true about the quantities in Columns I and II.

**A**  The number in Column I is greater.
**B**  The number in Column II is greater.
**C**  The two numbers are equal.
**D**  The relationship cannot be determined from the given information.

|     | Column I | | Column II | |
| --- | --- | --- | --- | --- |
| 10. | $5^3$ | | 15 | |
|     | Ⓐ   Ⓑ | | Ⓒ   Ⓓ | |
| 11. | $3^3 \cdot 3^2$ | | $3^6$ | |
|     | Ⓐ   Ⓑ | | Ⓒ   Ⓓ | |
| 12. | $10^{2+3}$ | | $(10^2)(10^3)$ | |
|     | Ⓐ   Ⓑ | | Ⓒ   Ⓓ | |
| 13. | $4^2$ | | $2^4$ | |
|     | Ⓐ   Ⓑ | | Ⓒ   Ⓓ | |

14. **Multiple Choice**  If $A = \frac{1}{2}bh$, what is $A$ when $b = 4a^2c$ and $h = 5a^3c^3$?
   - Ⓐ $10a^5c^2$
   - Ⓑ $10a^5c^4$
   - Ⓒ $20a^6c^3$
   - Ⓓ $20a^5c^4$

# Standardized Test Practice
## 8.2 Laws of Exponents: Powers and Products

**TEST TAKING STRATEGY** Use number sense to eliminate unreasonable choices.

1. **Multiple Choice** To what power is the 7 raised when $(7^2)^3$ is simplified?
   - Ⓐ 5
   - Ⓑ 6
   - Ⓒ 8
   - Ⓓ 3

2. **Multiple Choice** What is the value of the expression $(t^5)^3$ when $t = 2$?
   - Ⓐ 4
   - Ⓑ 1,024
   - Ⓒ 256
   - Ⓓ 32,768

3. **Multiple Choice** Which exponential expression is equivalent to 729?
   - Ⓐ $(3^3)^2$
   - Ⓑ $(4^2)^3$
   - Ⓒ $8^3$
   - Ⓓ $(9^2)^3$

4. **Multiple Choice** Which of the following is true when $x = 3$?
   - Ⓐ $(-x)^2 = 9$
   - Ⓑ $(-x)^2 = -9$
   - Ⓒ $(-x)^3 = -9$
   - Ⓓ $(-x)^3 = 27$

5. **Multiple Choice** When raising a power to a power, what operation is performed on the exponents?
   - Ⓐ addition
   - Ⓑ subtraction
   - Ⓒ multiplication
   - Ⓓ division

6. **Multiple Choice** What is 6 times 3 cubed?
   - Ⓐ 18
   - Ⓑ 324
   - Ⓒ 162
   - Ⓓ 648

7. **Multiple Choice** What is the coefficient of the expression $(3a^2b^3)^4$ if $a = 3$?
   - Ⓐ 12
   - Ⓑ 6561
   - Ⓒ 27
   - Ⓓ 531,441

**Quantitative Comparison** In Exercises 8–11, choose the letter of the statement below that is true about the quantities in Columns I and II.

A The number in Column I is greater.
B The number in Column II is greater.
C The two numbers are equal.
D The relationship cannot be determined from the given information.

|    | Column I | Column II |
|----|----------|-----------|
| 8. | $(-1)^3$ Ⓐ Ⓑ | $(-1)^2$ Ⓒ Ⓓ |
| 9. | $(a^3)^2$ Ⓐ Ⓑ | $(a^2)^3$ Ⓒ Ⓓ |
| 10. | $(14^3)^2$ Ⓐ Ⓑ | 100,000 Ⓒ Ⓓ |
| 11. | $-6x^3$ Ⓐ Ⓑ | $(-6x)^3$ Ⓒ Ⓓ |

12. **Multiple Choice** Which expression when evaluated for $x = 2$, is negative?
    - Ⓐ $-(-2x^2)^3$
    - Ⓑ $(-4x)^3$
    - Ⓒ $(-x^2)^2$
    - Ⓓ $-(-3x)^3$

13. **Multiple Choice** Which of the equations is true?
    - Ⓐ $(a^6b^3)^1 = a^7b^4$
    - Ⓑ $b^3 \cdot b^2 = b^6$
    - Ⓒ $(6a^2)^3 = 18a^6$
    - Ⓓ $(ab^2)^3 = a^3b^6$

Algebra 1

NAME _____ CLASS _____ DATE _____

# Standardized Test Practice
## 8.3 Laws of Exponents: Dividing Monomials

**TEST TAKING STRATEGY** Look at all of the answers before selecting one.

1. **Multiple Choice** What operation is performed on the exponents to simplify the expression $\frac{3^8}{3^3}$?
   - Ⓐ subtraction
   - Ⓑ multiplication
   - Ⓒ addition
   - Ⓓ division

2. **Multiple Choice** What is the value of $\frac{10^{12}}{10^{11}}$?
   - Ⓐ 100 $3t^5$
   - Ⓑ 10
   - Ⓒ 1000
   - Ⓓ $\frac{1}{10}$

3. **Multiple Choice** Which of the following does *not* equal $p^{10}$?
   - Ⓐ $p^8 \cdot p^2$
   - Ⓑ $\frac{p^{12}}{p^2}$
   - Ⓒ $(p^2)^5$
   - Ⓓ $\frac{p^{10}}{p}$

4. **Multiple Choice** If $\frac{4x^5}{2x^4} = 4$, find $x$.
   - Ⓐ 2
   - Ⓑ 9
   - Ⓒ 8
   - Ⓓ 0

5. **Multiple Choice** When the expression $\frac{-8x^3y^6}{4xy^3}$ is simplified, what is true about the exponent of $x$ compared to the exponent of $y$?
   - Ⓐ It will be equal to the $y$ exponent.
   - Ⓑ It stays the same.
   - Ⓒ It will be smaller.
   - Ⓓ It will be larger.

6. **Multiple Choice** If $s = 1$ and $t = 2$, what is the value of the expression $\frac{(3s^3t^2)^2(8st)}{12s^5t^3}$?
   - Ⓐ 0
   - Ⓑ 72
   - Ⓒ 24
   - Ⓓ 2730

7. **Multiple Choice** For which value of $t$ makes $(t^3)^2 = -64$ a true statement?
   - Ⓐ 3
   - Ⓑ 2
   - Ⓒ $-2$
   - Ⓓ none of the above

**Quantitative Comparison** In Exercises 8–11, choose the letter of the statement below that is true about the quantities in Columns I and II.

- **A** The number in Column I is greater.
- **B** The number in Column II is greater.
- **C** The two numbers are equal.
- **D** The relationship cannot be determined from the given information.

| | Column I | Column II | |
|---|---|---|---|
| 8. | $\frac{3^8}{3^6}$ | 9 | |
| | Ⓐ  Ⓑ | Ⓒ  Ⓓ | |
| 9. | $\frac{4^5}{4^3}$ | $\frac{5^4}{3^4}$ | |
| | Ⓐ  Ⓑ | Ⓒ  Ⓓ | |
| 10. | $8x^3$ | $\frac{80x^6}{10x^2}$ | |
| | Ⓐ  Ⓑ | Ⓒ  Ⓓ | |
| 11. | $\left(\frac{a^5b^2}{c^5}\right)$ | $\left(\frac{ab}{cd}\right)^3$ | |
| | Ⓐ  Ⓑ | Ⓒ  Ⓓ | |

NAME _____ CLASS _____ DATE _____

# Standardized Test Practice
## 8.4 Negative and Zero Exponents

**TEST TAKING STRATEGY** Substitute values in for variables to verify your answers.

**1. Multiple Choice** What is the value of $3^{-3}$?

- (A) 27
- (B) $-\dfrac{1}{27}$
- (C) $\dfrac{1}{27}$
- (D) $-9$

**2. Multiple Choice** If $\dfrac{6^4}{6^6} = 6^x$, what is the value of $x$?

- (A) $-2$
- (B) 2
- (C) 36
- (D) $\dfrac{1}{2}$

**3. Multiple Choice** What is $x^3 \cdot x^{-5}$ when $x = 4$?

- (A) $-16$
- (B) $-\dfrac{1}{16}$
- (C) 16
- (D) $\dfrac{1}{16}$

**4. Multiple Choice** What is true about any number raised to a zero power?

- (A) The number is 0.
- (B) The number is 1.
- (C) The number cannot be evaluated.
- (D) The number is insignificant.

**5. Multiple Choice** Which of the following is equivalent to $\dfrac{(5a^3)(2a^{-5})}{2a^{-4}}$?

- (A) $10a^2$
- (B) $10a^5$
- (C) $\dfrac{1}{5a^2}$
- (D) $5a^2$

**Quantitative Comparison** In Exercises 6–9, choose the letter of the statement below that is true about the quantities in Columns I and II.

- **A** The number in Column I is greater.
- **B** The number in Column II is greater.
- **C** The two numbers are equal.
- **D** The relationship cannot be determined from the given information.

| | Column I | Column II |
|---|---|---|
| 6. | 1 | $151^0$ |
| | (A) (B) | (C) (D) |
| 7. | $2^{-2}$ | $-\dfrac{1}{4}$ |
| | (A) (B) | (C) (D) |
| 8. | $\dfrac{15^8}{15^{12}}$ | $15^{-2}$ |
| | (A) (B) | (C) (D) |
| 9. | $t$ | $\dfrac{t^3 \cdot t^{-4}}{t^5}$ |
| | (A) (B) | (C) (D) |

**10. Multiple Choice** Which values complete the pattern in the table?

| 8 | 4 | 2 | | | | |
|---|---|---|---|---|---|---|
| $2^3$ | $2^2$ | $2^1$ | $2^0$ | $2^{-1}$ | $2^{-2}$ | $2^{-3}$ |

- (A) $0, \dfrac{1}{2}, \dfrac{1}{4}, \dfrac{1}{8}$
- (B) $0, -\dfrac{1}{2}, -\dfrac{1}{4}, -\dfrac{1}{8}$
- (C) $1, \dfrac{1}{2}, \dfrac{1}{4}, \dfrac{1}{8}$
- (D) $1, -\dfrac{1}{2}, -\dfrac{1}{4}, -\dfrac{1}{8}$

**11. Multiple Choice** What equals $\dfrac{1}{x^{-3}}$?

- (A) $-\dfrac{3}{x}$
- (B) $\dfrac{3}{x}$
- (C) $x^3$
- (D) $x^{-3}$

Algebra 1      Standardized Test Practice 8.4    53

NAME _____ CLASS _____ DATE _____

# Standardized Test Practice
## 8.5 Scientific Notation

**TEST TAKING STRATEGY**  Restate each question to make sure you have answered it correctly.

1. **Multiple Choice** How is the number 482,000 written using scientific notation?

    Ⓐ $4.82 \times 1000$  Ⓑ $4.82 \times 10^5$
    Ⓒ $4.82 \times 10^4$  Ⓓ $482 \times 10^5$

2. **Multiple Choice** How many digits are in the number $7.5 \times 10^{11}$?

    Ⓐ 11  Ⓑ 12
    Ⓒ 13  Ⓓ 14

3. **Multiple Choice** Numbers written in scientific notation are _____.

    Ⓐ expanded
    Ⓑ used only in science
    Ⓒ abbreviated
    Ⓓ not significant

4. **Multiple Choice** If $0.00000000831 = 8.31 \times 10^x$, what is the value of $x$?

    Ⓐ $-8$  Ⓑ $-9$
    Ⓒ $9$   Ⓓ $10$

5. **Multiple Choice** Which number correctly expresses $4.65 \times 10^5$ in decimal notation?

    Ⓐ 46,500     Ⓑ 465,000
    Ⓒ 4,650,000  Ⓓ 46,500,000

6. **Multiple Choice** When written in decimal notation, how many zeros does $(3 \times 10^6)(4 \times 10^3)$ have?

    Ⓐ 9  Ⓑ 18
    Ⓒ 6  Ⓓ 3

7. **Multiple Choice** What is the value of $y$ in the following equation?
   $(1.8 \times 10^y)(1.2 \times 10^8) = 2.16 \times 10^4$

    Ⓐ 4   Ⓑ 64
    Ⓒ $-2$  Ⓓ $-4$

*Quantitative Comparison* In Exercises 8–11, choose the letter of the statement below that is true about the quantities in Columns I and II.

**A** The number in Column I is greater.
**B** The number in Column II is greater.
**C** The two numbers are equal.
**D** The relationship cannot be determined from the given information.

|     | Column I | Column II |
|-----|----------|-----------|
| 8.  | $1.3 \times 10^3$ | 1300 |
|     | Ⓐ  Ⓑ  Ⓒ  Ⓓ | |
| 9.  | $7.7 \times 10^{-6}$ | $-7.7 \times 10^6$ |
|     | Ⓐ  Ⓑ  Ⓒ  Ⓓ | |
| 10. | $5.2 \times 10^{-1}$ | $(1.3 \times 10^{-3})(4 \times 10^4)$ |
|     | Ⓐ  Ⓑ  Ⓒ  Ⓓ | |
| 11. | $\dfrac{7 \times 10^7}{14 \times 10^9}$ | $\dfrac{1}{2}$ |
|     | Ⓐ  Ⓑ  Ⓒ  Ⓓ | |

12. **Multiple Choice** Sam is 11 years old today. Which of the following shows his age in minutes?

    Ⓐ $6.9 \times 10^9$   Ⓑ $5.78 \times 10^6$
    Ⓒ $9.6 \times 10^5$   Ⓓ $4.0 \times 10^8$

13. **Multiple Choice** In 1999, the world's largest bank had net assets of $6,919,203,000,000. What is this amount in terms using scientific notation?

    Ⓐ $6.9 \times 10^{12}$    Ⓑ $69,129 \times 10^{12}$
    Ⓒ $6.9 \times 10^{-12}$   Ⓓ $0.69 \times 10^{12}$

NAME _____  CLASS _____  DATE _____

# Standardized Test Practice
## 8.6 Exponential Functions

**TEST TAKING STRATEGY** Look for math ideas that are presented in words.

1. **Multiple Choice** Exponential functions are functions in which the values in the range are changed by a fixed ____.
   - (A) *x*-value
   - (B) amount
   - (C) rate
   - (D) exponent

2. **Multiple Choice** The graphs of exponential functions are *never*
   - (A) linear.
   - (B) able to be graphed.
   - (C) in Quadrant IV.
   - (D) curved lines.

3. **Multiple Choice** What is the principal amount for the following investment?
   $$P = 2000(1 + 5\%)^2$$
   - (A) $2205
   - (B) $4500
   - (C) $6000
   - (D) $2020.05

4. **Multiple Choice** What is the value of *A* for the expression $2280 = A(1 + 14\%)$?
   - (A) $2205
   - (B) $748
   - (C) $1456
   - (D) $2000

5. **Multiple Choice** In 1999, the world's population was about 6 billion and growing at a rate of 1.4%. Estimate the world's population after 5 years.
   - (A) 30.52 billion
   - (B) 6.43 billion
   - (C) 25.7 billion
   - (D) 6.34 billion

6. **Multiple Choice** The population of a small country was about 250 million in 1998 and growing at a rate of 0.6%. About how much will the population increase in 10 years?
   - (A) 15.4 million
   - (B) 64.8 million
   - (C) 6.36 million
   - (D) 447.5 million

7. **Multiple Choice** Sally's retirement fund has an interest rate of 18%. After 3 years her balance was $6250. To the nearest dollar, what was her original investment?
   - (A) $3804
   - (B) $3839
   - (C) $6859
   - (D) $10,269

**Quantitative Comparison** In Exercises 8–11, choose the letter of the statement below that is true about the quantities in Columns I and II.

**A** The number in Column I is greater.
**B** The number in Column II is greater.
**C** The two numbers are equal.
**D** The relationship cannot be determined from the given information.

Use the function $f(x) = 2x^{2x}$ for Exercises 8–11.

| | Column I | Column II |
|---|---|---|
| 8. | $f(2)$ | 4 |
| | (A) (B) | (C) (D) |
| 9. | $f(-3)$ | $-0.15$ |
| | (A) (B) | (C) (D) |
| 10. | $f(0)$ | 0 |
| | (A) (B) | (C) (D) |
| 11. | $f(-1)$ | 0.25 |
| | (A) (B) | (C) (D) |

12. **Multiple Choice** What is the missing value in the table for the function $y = 1.5^x$?

| $x$ | $-2$ | $-1$ | 0 | 1 | 2 |
|---|---|---|---|---|---|
| $1.5^x$ | $0.44\overline{4}$ | ? | 1 | 1.5 | 2.25 |

   - (A) $-6.\overline{6}$
   - (B) $-2.25$
   - (C) 66
   - (D) $0.6\overline{6}$

Algebra 1  Standardized Test Practice 8.6  55

NAME _____ CLASS _____ DATE _____

# Standardized Test Practice
## 8.7 Applications of Exponential Functions

**TEST TAKING STRATEGY** Use an estimate to check if your answer is reasonable.

1. **Multiple Choice** Which real-world situation can *not* be modeled by exponential functions?

   Ⓐ the value of an antique chair
   Ⓑ the age of a vintage car
   Ⓒ the age of a fossil
   Ⓓ the population of a city in 40 years

2. **Multiple Choice** To determine exponential decay, the $r$ in the formula $P = A(1 + r)^t$ is:

   Ⓐ positive    Ⓑ squared
   Ⓒ negative    Ⓓ divided by 2

3. **Multiple Choice** The value of a limited edition Corvette has been growing at a rate of 5% per year for the past 3 years. If the original cost of the car was $31,500, what is it worth today?

   Ⓐ $36,465    Ⓑ $106,312
   Ⓒ $99,225    Ⓓ $567,000

4. **Multiple Choice** Dominic's parents want to have $4500 for tuition saved before he goes to college in 4 years. The interest rate on the college fund is 6.5% and is compounded annually. How much does their original deposit need to be in order to meet their goal in 4 years?

   Ⓐ $1002    Ⓑ $3498
   Ⓒ $1500    Ⓓ $3730

5. **Multiple Choice** Janice knows she can expect to get a given step in an algebra problem correct 95% of the time. She does 100 problems, each with 4 steps. In how many problems can she expect to get all 4 steps correct?

   Ⓐ 0.81    Ⓑ 80
   Ⓒ 81      Ⓓ 82

6. **Multiple Choice** The population of a small city has been declining slowly at a rate of 2% a year. Five years ago the population was 13,565. Which is the best estimate of the population of the city today?

   Ⓐ 13,290    Ⓑ 12,260
   Ⓒ 4340      Ⓓ 13,200

7. **Multiple Choice** Use carbon dating to determine the age of a fossil with 40% of carbon-14 remaining.

   Ⓐ 5000 years    Ⓑ 10,000 years
   Ⓒ 18,000 years  Ⓓ 25,000 years

*Quantitative Comparison* In Exercises 8–10, choose the letter of the statement below that is true about the quantities in Columns I and II.

**A** The number in Column I is greater.
**B** The number in Column II is greater.
**C** The two numbers are equal.
**D** The relationship cannot be determined from the given information.

| | Column I | Column II |
|---|---|---|
| 8. | population of a city in 2 years with 2% growth and 15,000 people | population of a city in 3 years with 4% decline and 18,000 people |
| | Ⓐ  Ⓑ | Ⓒ  Ⓓ |
| 9. | $6500 with 9% annual interest in 4 years | $9500 |
| | Ⓐ  Ⓑ | Ⓒ  Ⓓ |
| 10. | $6\left(\dfrac{1}{10}\right)^3$ | 0.006 |
| | Ⓐ  Ⓑ | Ⓒ  Ⓓ |

# SAT/ACT Chapter Test

## Chapter 8  Exponents and Exponential Functions

**TEST TAKING STRATEGY**  Be sure you understand what is being asked in the question.

1. **Multiple Choice** Which value is equal to $7^5$?
   - (A) 35
   - (B) 175
   - (C) 4375
   - (D) 16,807

2. **Multiple Choice** Simplify $(3x^4y^2z^6)(8y^3z^5)$.
   - (A) $11x^4y^5z^{11}$
   - (B) $24x^4y^5z^{11}$
   - (C) $24x^4y^6z^{30}$
   - (D) $11y^5z^{11}$

**Quantitative Comparison**  In Exercises 3–6, choose the letter of the statement below that is true about the quantities in Columns I and II.

**A** The number in Column I is greater.
**B** The number in Column II is greater.
**C** The two numbers are equal.
**D** The relationship cannot be determined from the given information.

| | Column I | Column II |
|---|---|---|
| 3. | $3^{-4}$ | $4^{-3}$ |
|  | (A)  (B) | (C)  (D) |
| 4. | $7.34 \times 10^{-6}$ | $-7.34 \times 10^6$ |
|  | (A)  (B) | (C)  (D) |
| 5. | $(4.5 \times 10^{-3})^0$ | 0 |
|  | (A)  (B) | (C)  (D) |
| 6. | $(3 \times 10^3)(6 \times 10^{-7})$ | 0.0018 |
|  | (A)  (B) | (C)  (D) |

7. **Multiple Choice** When you simplify $(15a^3b^2)^3$, what is the exponent of $b$?
   - (A) $b^2$
   - (B) $b^5$
   - (C) $b^4$
   - (D) $b^6$

8. **Multiple Choice** Using exponent rules, which of the following is equivalent to $\frac{16^{12}}{16^3}$?
   - (A) $16^9$
   - (B) $16^4$
   - (C) $16^{36}$
   - (D) 144

9. **Multiple Choice** Which expression is the simplified form of $\frac{(3t^2s)(4t^3s^2)}{48t^5s^5}$?
   - (A) $\frac{1}{4s^2}$
   - (B) 0
   - (C) $\frac{t}{4s^3}$
   - (D) $\frac{-s^2}{4t}$

10. **Multiple Choice** In the last 3 years, Steve's investments averaged an annual interest rate of 19%. If his original investment was $500, what is this investment worth now?
    - (A) $1785
    - (B) $529.04
    - (C) $842.58
    - (D) $3420

11. **Multiple Choice** The population of a town has been declining at a rate of 3% in the last 3 years. If the population today is 19,875, what was the population 3 years ago?
    - (A) 21,777
    - (B) 18,139
    - (C) 20,874
    - (D) 21,840

12. **Multiple Choice** The population of a small town is 150,000 and is growing at a 2% rate. When will the population be over 175,000?
    - (A) 3 years
    - (B) 8 years
    - (C) 5 years
    - (D) 25 years

NAME _____ CLASS _____ DATE _____

# Standardized Test Practice
## 9.1 Adding and Subtracting Polynomials

**TEST TAKING STRATEGY** Common mistakes are usually included in the answer choices.

1. **Multiple Choice** Which expression represents $x^2 + 6 - 4x + 3x^2$ in standard form?
   - Ⓐ $x^2 + 4x - 6$
   - Ⓑ $4x^2 - 4x - 6$
   - Ⓒ $2x^2 - 4x + 6$
   - Ⓓ $4x^2 - 4x + 6$

2. **Multiple Choice** What is the sum of $(x^2 + 8 - 2x)$ and $(6x^2 + 4x - 3)$?
   - Ⓐ $7x^2 + 12x - 5x$
   - Ⓑ $7x^2 + 5 + 2x$
   - Ⓒ $7x^2 + 2x + 5$
   - Ⓓ $7x^2 - 2x + 5$

3. **Multiple Choice** What is $(7x + -4x^2 + 10)$ less $(13x - x^2 + 2)$?
   - Ⓐ $3x^2 + 6x + 8$
   - Ⓑ $-3x^2 - 6x + 8$
   - Ⓒ $-20x + 5x^2 + 12$
   - Ⓓ $-5x^2 - 6x + 12$

4. **Multiple Choice** What is the difference of $(12x^3 - 2x)$ from $(-6x^3 + 3x)$?
   - Ⓐ $-18x^3 - 2$
   - Ⓑ $6x^3 - x$
   - Ⓒ $-4x^3$
   - Ⓓ $-18x^3 + 5x$

5. **Multiple Choice** What type of polynomial is $4x^3 + 6x^2 + 2$?
   - Ⓐ quadratic trinomial
   - Ⓑ cubic polynomial
   - Ⓒ cubic trinomial
   - Ⓓ linear polynomial

6. **Multiple Choice** Which polynomial would be defined as a quadratic binomial?
   - Ⓐ $4x^2 + 2$
   - Ⓑ $8x^3 + 12$
   - Ⓒ $4x^3 + 2$
   - Ⓓ $8x^4 + 4x^2$

7. **Multiple Choice** What is the perimeter of a rectangle with a length of $(3x^2 + 2)$ and a width of $(4x + 1)$?
   - Ⓐ $3x^2 + 4x + 3$
   - Ⓑ $14x^2 + 3$
   - Ⓒ $6x^2 + 8x + 6$
   - Ⓓ $10x^2 + 4x + 3$

8. **Multiple Choice** What is the missing term?

   $15x^2 - 26x + 1 - (-30x^2 - 11x + 1)$
   $= 45x^2 - \underline{\phantom{xx}}$
   - Ⓐ $37x$
   - Ⓑ $-37x$
   - Ⓒ $15x$
   - Ⓓ $-15x$

*Quantitative Comparison* In Exercises 9–11, choose the letter of the statement below that is true about the quantities in Columns I and II.

**A** The number in Column I is greater.
**B** The number in Column II is greater.
**C** The two numbers are equal.
**D** The relationship cannot be determined from the given information.

| | Column I | Column II |
|---|---|---|
| 9. | the number of terms in a quartic trinomial | the number of terms in a cubic binomial |
| | Ⓐ  Ⓑ | Ⓒ  Ⓓ |
| 10. | the degree of a constant | the degree of a linear monomial |
| | Ⓐ  Ⓑ | Ⓒ  Ⓓ |
| 11. | the perimeter of a square with sides of $(7x + 2)$ when $x = 4$ | the perimeter of an equilateral triangle with sides of $(10x + 1)$ when $x = 3$ |
| | Ⓐ  Ⓑ | Ⓒ  Ⓓ |

# Standardized Test Practice
## 9.2 Modeling Polynomial Multiplication

**TEST TAKING STRATEGY** Substitute values in for variables to verify your answers.

**1. Multiple Choice** What are the binomial factors in the model below?

[model diagram]

- Ⓐ $(2x + 2x)(2x)$
- Ⓑ $(2x + 2)(x + 1)$
- Ⓒ $(3x) + (3)$
- Ⓓ $(x + 1)(x^2 + 2)$

**2. Multiple Choice** What trinomial does the algebra tile model represent?

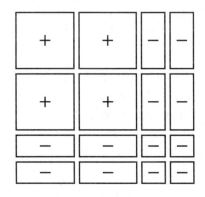

- Ⓐ $4x^2 + 8x - 4$
- Ⓑ $4x^2 - 8x - 4$
- Ⓒ $4x^2 - 4x + 4$
- Ⓓ $8x^2 - 8x$

**3. Multiple Choice** Which set of factors equals $a^2 - b^2$?

- Ⓐ $(a - b)(a - b)$
- Ⓑ $(a - b)(a + b)$
- Ⓒ $(a^2 - a)(b^2 - b)$
- Ⓓ $(a^2 - b)(a - b)$

**4. Multiple Choice** Using the rules for special products, what is the product of $(x + 3)(x + 3)$?

- Ⓐ $2x^2 + 6x + 6$
- Ⓑ $2x^2 + 6x + 9$
- Ⓒ $x^2 + 6x + 6$
- Ⓓ $x^2 + 6x + 9$

**5. Multiple Choice** What are the two factors of $x^2 - 10x + 25$?

- Ⓐ $(x - 5)(x + 5)$
- Ⓑ $-(x + 5)^2$
- Ⓒ $(x - 5)^2$
- Ⓓ $(x - 5)(x - 2)$

**6. Multiple Choice** What is the product of $6(x + 7)$?

- Ⓐ $6x + 13$
- Ⓑ $6x + 7$
- Ⓒ $6x + 42$
- Ⓓ $x - 42$

**7. Multiple Choice** Which is *not* a factor of $(-3a^2 + 6a)$?

- Ⓐ $3$
- Ⓑ $6a$
- Ⓒ $-3a$
- Ⓓ $a$

***Quantitative Comparison*** In Exercises 8–11, choose the letter of the statement below that is true about the quantities in Columns I and II.

**A** The number in Column I is greater.
**B** The number in Column II is greater.
**C** The two numbers are equal.
**D** The relationship cannot be determined from the given information.

| | Column I | | Column II | |
|---|---|---|---|---|
| 8. | $(x - 5)^2$ | | $(x - 5)(x - 5)$ | |
| | Ⓐ | Ⓑ | Ⓒ | Ⓓ |
| 9. | $3(4 + 9)$ | | $12 + 27$ | |
| | Ⓐ | Ⓑ | Ⓒ | Ⓓ |
| 10. | the area of a square with 4-cm sides | | 16 cm² | |
| | Ⓐ | Ⓑ | Ⓒ | Ⓓ |
| 11. | $6(x + 2)$ | | $2(x + 6)$ | |
| | Ⓐ | Ⓑ | Ⓒ | Ⓓ |

NAME _____ CLASS _____ DATE _____

# Standardized Test Practice
## 9.3 Multiplying Binomials

**TEST TAKING STRATEGY** Be aware of similar answers.

1. **Multiple Choice** Using the Distributive Property, what is the following product?

   $$2x(4x + 3)$$

   Ⓐ $14x^2$  Ⓑ $8x^2 + 6x$
   Ⓒ $8x^2 + 3x$  Ⓓ $4x^2 + 2x + 3$

2. **Multiple Choice** If you use the Distributive Property to multiply the binomial $(x + 4)(x - 6)$, you will get:

   Ⓐ $x(x - 6) + 4(x - 6)$
   Ⓑ $(2x^2 + 4x) - 6$
   Ⓒ $4x(x - 6)$
   Ⓓ $(x + x)(4 - 6)$

3. **Multiple Choice** In the FOIL method of multiplying binomials, which two terms would be the "I" in the factors?

   $$(x - 3)(x - 5)$$

   Ⓐ $x, x$  Ⓑ $x, 5$
   Ⓒ $-3, x$  Ⓓ $-3, 5$

4. **Multiple Choice** What is the product of the two factors that would be multiplied together in the fourth step of the FOIL method?

   $$(x + 7)(-x + 2)$$

   Ⓐ $-x^2$  Ⓑ $-7x$
   Ⓒ $2x$  Ⓓ $14$

5. **Multiple Choice** In the following equation, what is the value of $x$?

   $$(x + 2)^2 - (x - 1)^2 = 15$$

   Ⓐ $1$  Ⓑ $2$
   Ⓒ $3$  Ⓓ $4$

6. **Multiple Choice** The area of the shaded part of the figure is 25 square centimeters. What is the length of the larger rectangle?

   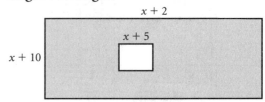

   Ⓐ 17 centimeters  Ⓑ 25 centimeters
   Ⓒ 20 centimeters  Ⓓ 12 centimeters

**Quantitative Comparison** In Exercises 7–9, choose the letter of the statement below that is true about the quantities in Columns I and II.

A The number in Column I is greater.
B The number in Column II is greater.
C The two numbers are equal.
D The relationship cannot be determined from the given information.

| | Column I | Column II |
|---|---|---|
| 7. | area of a square with sides $x + 2$ when $x = 5$ | the area of a rectangle with a length of $x$ and a width of $x + 4$ when $x = 4$ |
| | Ⓐ  Ⓑ | Ⓒ  Ⓓ |
| 8. | the last term in the product $(x - 3)^2$ | $-9$ |
| | Ⓐ  Ⓑ | Ⓒ  Ⓓ |
| 9. | the area, in terms of $y$, of a square frame with sides of $(y - 2)$ | the area, in terms of $y$, of a square with sides of $(y + 2)$ |
| | Ⓐ  Ⓑ | Ⓒ  Ⓓ |

NAME _____ CLASS _____ DATE _____

# Standardized Test Practice
## 9.4 Polynomial Functions

**TEST TAKING STRATEGY** Be sure you understand what is being asked in the question.

1. **Multiple Choice** If a rectangular prism has a length of 4 centimeters, a width of 8 centimeters and a height of 10 centimeters, what is the volume?
   - Ⓐ 320 cm
   - Ⓑ 320 cm$^2$
   - Ⓒ 320 cm$^3$
   - Ⓓ 3200 cm$^3$

2. **Multiple Choice** Which equation represents the surface area of a solid with a height of 15 centimeters, a width of $x$, and a length of 30 centimeters?
   - Ⓐ $(45x + 900)$cm$^2$
   - Ⓑ $(90x + 900)$cm$^2$
   - Ⓒ $(9x + 900)$cm$^2$
   - Ⓓ $40,500x$ cm$^2$

3. **Multiple Choice** An equation that is true for all values or variables is called:
   - Ⓐ a function.
   - Ⓑ an identity.
   - Ⓒ an equation.
   - Ⓓ absolute value.

4. **Multiple Choice** Which value of $x$ makes the following equation true?
   $$x^2 - 16 = (x - 4)(x + 4)$$
   - Ⓐ 0
   - Ⓑ 1
   - Ⓒ 2
   - Ⓓ all of the above

5. **Multiple Choice** Which is *not* an example of a polynomial function?
   - Ⓐ $g(x) = x^2 + 2$
   - Ⓑ $g(x) = \sqrt{x} + 3$
   - Ⓒ $g(x) = x^2 + 3x + 2$
   - Ⓓ $g(x) = 2x^2 + 1$

6. **Multiple Choice** Which polynomial function is true for $x$-values $-1, 0$ and $1$?
   - Ⓐ $x^2 + 3x + 2 = (x + 2)(x + 1)$
   - Ⓑ $x^2 + 36 = (x + 6)(x + 6)$
   - Ⓒ $x^2 - 4x + 32 = (x + 4)(x - 8)$
   - Ⓓ $x^2 + 15x + 50 = (-x + 5)(x + 10)$

7. **Multiple Choice** Which equation is an identity?
   - Ⓐ $4x^2 - 16 = (2x - 4)(2x + 4)$
   - Ⓑ $x^3 + x^2 + x + 1 = (x + 2)^3$
   - Ⓒ $3x^2 - 16 = (3x - 4)(x + 4)$
   - Ⓓ $x^2 - 7x - 14 = (x + 2)(x - 7)$

*Quantitative Comparison* In Exercises 8–11, choose the letter of the statement below that is true about the quantities in Columns I and II.

**A** The number in Column I is greater.
**B** The number in Column II is greater.
**C** The two numbers are equal.
**D** The relationship cannot be determined from the given information.

| | Column I | Column II |
|---|---|---|
| 8. | volume of a cube with 3-inch sides | volume of a 2 inch × 1 inch × 3 inch rectangular prism |
| | Ⓐ Ⓑ | Ⓒ Ⓓ |
| 9. | volume of a cylinder with a height of 8 meters and a radius of $x$ meters | volume of a cylinder with a height of 4 meters and a radius of $2x$ meters |
| | Ⓐ Ⓑ | Ⓒ Ⓓ |
| 10. | $y = (x + 4)^3$ for $x = -1$ | $y = x^3 + 16x^2 + 48x + 64$ for $x = -1$ |
| | Ⓐ Ⓑ | Ⓒ Ⓓ |
| 11. | the surface area of a cube with 6-centimeter sides | 18 cm$^2$ |
| | Ⓐ Ⓑ | Ⓒ Ⓓ |

Algebra 1      Standardized Test Practice 9.4

NAME _____ CLASS _____ DATE _____

# Standardized Test Practice
## 9.5 Common Factors

**TEST TAKING STRATEGY** Look for obvious distractors and eliminate them as answer choices.

1. **Multiple Choice** Which polynomial can *not* be factored?
   - Ⓐ $x^2 - 3$
   - Ⓑ $12x^3 - 4$
   - Ⓒ $8x - 4$
   - Ⓓ $10y^3 + 5y$

2. **Multiple Choice** What is the greatest common factor of $12x^3 + 6x$?
   - Ⓐ $2x$
   - Ⓑ $3x$
   - Ⓒ $6x$
   - Ⓓ $6x^2$

3. **Multiple Choice** Which polynomial has a greatest common factor of $5x$?
   - Ⓐ $25x^3 + 5x$
   - Ⓑ $5x - 5$
   - Ⓒ $15x^2 + 5$
   - Ⓓ $30x^2 - 10$

4. **Multiple Choice** The following is an example of what type of factoring?
   $2x^2 + x + 6x + 3 =$
   $(2x^2 + x) + (6x + 3)$
   - Ⓐ grouping
   - Ⓑ binomial factoring
   - Ⓒ GCF
   - Ⓓ LCD

5. **Multiple Choice** What are the factors of $x(x + 3) - 4(x + 3)$?
   - Ⓐ $(x + 3)^2 + (x + 4)$
   - Ⓑ $x^2 + 3x - 4x - 12$
   - Ⓒ $(2 - 4)(x + 3)^2$
   - Ⓓ $(x - 4)(x + 3)$

6. **Multiple Choice** What is the binomial factor in the following polynomial?
   $4s(w - 6) + 4(w - 6)$
   - Ⓐ $w$
   - Ⓑ $w - 6$
   - Ⓒ $4s$
   - Ⓓ $4s + 4$

**Quantitative Comparison** In Exercises 7–10, choose the letter of the statement below that is true about the quantities in Columns I and II.

**A** The number in Column I is greater.
**B** The number in Column II is greater.
**C** The two numbers are equal.
**D** The relationship cannot be determined from the given information.

| | Column I | Column II |
|---|---|---|
| 7. | $7ab^2 - 21a^3b$ | $7ab(b - 3a^2)$ |
| | Ⓐ Ⓑ | Ⓒ Ⓓ |
| 8. | the greatest common factor of 21 and 56 | the greatest common factor of 8 and 12 |
| | Ⓐ Ⓑ | Ⓒ Ⓓ |
| 9. | The value of $A$ in $A = \pi R^2 - \pi r^2$, where $R = 4$ and $r = 2$. | $10\pi$ |
| | Ⓐ Ⓑ | Ⓒ Ⓓ |
| 10. | the area of a square with sides of $2x + 4$ | the area of a square with sides of $2(x + 1)$ |
| | Ⓐ Ⓑ | Ⓒ Ⓓ |

11. **Multiple Choice** What is the factored form of $16(x - 3) - y(x - 3)$?
    - Ⓐ $(16 - y)(x - 3)$
    - Ⓑ $(16 + y)(x - 3)$
    - Ⓒ $(x + 3)^2(16 + y)$
    - Ⓓ $(16 - y) - (x - 3)$

12. **Multiple Choice** Which polynomial can be factored to $4a^2b(4a + 5b^2)$?
    - Ⓐ $16a^3 + 20b^3$
    - Ⓑ $16a^3b + 20a^2b^3$
    - Ⓒ $16a^3b + 5b^2$
    - Ⓓ $16a^3b + 20ab^3$

# Standardized Test Practice
## 9.6 Factoring Special Polynomials

**TEST TAKING STRATEGY** Read the question again to be sure it is answered completely.

1. **Multiple Choice** Which is *not* an example of a perfect square trinomial?
   - Ⓐ $25x + 10x + 4$
   - Ⓑ $16x^2 + 24x + 9$
   - Ⓒ $4x^2 + 8x + 4$
   - Ⓓ $x^2 - 8x + 16$

2. **Multiple Choice** What is *not* a true statement about perfect square trinomials?
   - Ⓐ The first term is a perfect square.
   - Ⓑ The second term is twice the product of the first and last terms.
   - Ⓒ All of the terms are positive.
   - Ⓓ The last term is a perfect square.

3. **Multiple Choice** Which expression can be factored using the difference of two squares?
   - Ⓐ $6x^2 - 16$
   - Ⓑ $9x^2 + 4$
   - Ⓒ $2x^2 - 25$
   - Ⓓ $4x^2 - 9$

**Quantitative Comparison** In Exercises 4–5, choose the letter of the statement below that is true about the quantities in Columns I and II.

- **A** The number in Column I is greater.
- **B** The number in Column II is greater.
- **C** The two numbers are equal.
- **D** The relationship cannot be determined from the given information.

| | Column I | Column II | | |
|---|---|---|---|---|
| 4. | $18 \cdot 22$ | $(20 + 2)(20 - 2)$ | | |
| | Ⓐ  Ⓑ | Ⓒ  Ⓓ | | |
| 5. | $37 \cdot 43$ | $40^2 - 4^2$ | | |
| | Ⓐ  Ⓑ | Ⓒ  Ⓓ | | |

6. **Multiple Choice** The area of a square is $x^2 - 14x + 49$. What is the perimeter of the square?
   - Ⓐ $4x + 28$
   - Ⓑ $x + 7$
   - Ⓒ $x^2 + 2408$
   - Ⓓ $4x - 28$

7. **Multiple Choice** Which of the following is equivalent to $(s^2 - t^2)$?
   - Ⓐ $(s - t)(s + t)$
   - Ⓑ $(s - t)(s - t)$
   - Ⓒ $(s - t)^2$
   - Ⓓ $st(s - t)$

8. **Multiple Choice** What is the length of each side of a square that has an area of $25x^2 - 90x + 81$?
   - Ⓐ $x - 9$
   - Ⓑ $5x - 9$
   - Ⓒ $5x + 9$
   - Ⓓ $5(x - 9)$

9. **Multiple Choice** What are the factors of the following expression?

$$y^2 - 144$$

   - Ⓐ $(y + 12) + (y - 12)$
   - Ⓑ $(y^2 - 12)(y - 12)$
   - Ⓒ $(y - 12)(y + 12)$
   - Ⓓ $(y - 72)(y + 72)$

10. **Multiple Choice** When factoring a polynomial using the difference of two squares, what is true about the factors?
    - Ⓐ Both factors are negative.
    - Ⓑ Only one term in each factor is a square root.
    - Ⓒ Only one term in one of the factors can be negative.
    - Ⓓ none of the above

Algebra 1

# Standardized Test Practice
## 9.7 Factoring Quadratic Trinomials

**TEST TAKING STRATEGY** Make an inference to fill in unknown or missing information.

1. **Multiple Choice** If the third term of a trinomial is 24, which of the following can *not* be its factors?
   - Ⓐ $(x + 4)(x - 6)$
   - Ⓑ $(x + 3)(8 + x)$
   - Ⓒ $(x - 2)(x - 12)$
   - Ⓓ $(x - 4)(x - 6)$

2. **Multiple Choice** If the area of a rectangle is $x^2 - 8x + 15$, what factors represent the lengths of the sides?
   - Ⓐ $(x - 7)(x - 1)$
   - Ⓑ $(x + 3)(x - 5)$
   - Ⓒ $(x - 15)(x + 7)$
   - Ⓓ $(x - 3)(x - 5)$

3. **Multiple Choice** What trinomial is produced by the factors $(3x)(x - 5)(x + 1)$?
   - Ⓐ $3x^2 - 12x - 15$
   - Ⓑ $3x^3 - 12x^2 - 15x$
   - Ⓒ $3x^4 - 5x$
   - Ⓓ $3x^4 - 12x^2 - 15$

*Quantitative Comparison* In Exercises 4–5, choose the letter of the statement below that is true about the quantities in Columns I and II.

A The number in Column I is greater.
B The number in Column II is greater.
C The two numbers are equal.
D The relationship cannot be determined from the given information.

|   | Column I | Column II |   |
|---|---|---|---|
| 4. | GCF of $15a^3 + 18a - 60$ | GCF of $21a^3 + 7$ |   |
|   | Ⓐ   Ⓑ | Ⓒ   Ⓓ |   |
| 5. | the number of factors of 15 | the number of factors of 36 |   |
|   | Ⓐ   Ⓑ | Ⓒ   Ⓓ |   |

6. **Multiple Choice** What represents the dimensions of a quadrilateral with an area of $25x^2 - 121$?
   - Ⓐ $(5x - 11)(5x - 11)$
   - Ⓑ $(5x + 11)(5x + 11)$
   - Ⓒ $(5x - 11)(5x + 11)$
   - Ⓓ $x(11 + 5x)(11 - 5x)$

7. **Multiple Choice** How many possible values of $b$ are there for $x^2 + bx - 16$?
   - Ⓐ 3
   - Ⓑ 4
   - Ⓒ 5
   - Ⓓ 6

8. **Multiple Choice** Which can *not* be a possible value for $b$ in the trinomial $x^2 + bx - 16$?
   - Ⓐ $-15$
   - Ⓑ $-10$
   - Ⓒ 6
   - Ⓓ 0

9. **Multiple Choice** What are the signs of the constant terms of the binomial factors of the polynomial $ax^2 - bx + c$?
   - Ⓐ Both are positive.
   - Ⓑ Both are negative.
   - Ⓒ One is positive.
   - Ⓓ One is negative.

10. **Multiple Choice** Which of the following trinomials is prime?
    - Ⓐ $x^2 + 15x - 100$
    - Ⓑ $a^2 + 3a - 6$
    - Ⓒ $y^2 - 13y - 48$
    - Ⓓ $n^2 + 7n + 12$

11. **Multiple Choice** What are the factors of $x^2 + 15x - 100$?
    - Ⓐ $(x + 100)(x - 85)$
    - Ⓑ $(x - 10)(x + 10)$
    - Ⓒ $(x + 25)(x - 4)$
    - Ⓓ $(x + 20)(x - 5)$

NAME _____ CLASS _____ DATE _____

# Standardized Test Practice
## 9.8 Solving Equations by Factoring

**TEST TAKING STRATEGY** Answer the questions that are the easiest first.

1. **Multiple Choice** What are the zeros of the function $y = x^2 - 7x + 12$?
   - Ⓐ 3, 4
   - Ⓑ 0, 0
   - Ⓒ −3, 4
   - Ⓓ −4, −3

2. **Multiple Choice** What is the solution to the equation $18x^2 - 44x = 0$?
   - Ⓐ 0, 0
   - Ⓑ 0, 2
   - Ⓒ 0, 3
   - Ⓓ none of the above

3. **Multiple Choice** Which polynomial function has zeros of 6 and −5?
   - Ⓐ $x^2 - 6x - 30$
   - Ⓑ $x^2 - x - 30$
   - Ⓒ $x^2 + 5x - 30$
   - Ⓓ $x^2 + x - 30$

4. **Multiple Choice** What is the solution to the equation $0 = x^2 + 9x + 8$?
   - Ⓐ −8, −1
   - Ⓑ 8, −1
   - Ⓒ 8, 1
   - Ⓓ −8, 1

5. **Multiple Choice** What do the zeros represent in a function that models the height of a projectile over time?
   - Ⓐ the maximum height of the projectile
   - Ⓑ the time when the projectile is at the ground
   - Ⓒ how fast the projectile moves
   - Ⓓ how fast the projectile falls

6. **Multiple Choice** The length of a rectangle is 10 meters more than its width. What is the length of the rectangle if the area is 144 square meters?
   - Ⓐ 18 meters
   - Ⓑ 8 meters
   - Ⓒ 28 meters
   - Ⓓ 12 meters

*Quantitative Comparison* In Exercises 7–10, choose the letter of the statement below that is true about the quantities in Columns I and II.

A  The number in Column I is greater.
B  The number in Column II is greater.
C  The two numbers are equal.
D  The relationship cannot be determined from the given information.

| | Column I | Column II |
|---|---|---|
| 7. | the length of a side of a square that has an area of 25 cm² | 5 cm |
| | Ⓐ  Ⓑ  Ⓒ  Ⓓ | |
| 8. | the value of $-36x^2 + 18x - 1$ when $x = 0$ | 1 |
| | Ⓐ  Ⓑ  Ⓒ  Ⓓ | |
| 9. | the area of a garden with a perimeter of 200 feet and a length of 30 feet | the area of a garden with a perimeter of 200 feet and a length of 50 feet |
| | Ⓐ  Ⓑ  Ⓒ  Ⓓ | |
| 10. | the perimeter of a rectangular pool with an area of 150 square meters and a length of 30 meters | the perimeter of a rectangular pool with an area of 100 square meters and a length of 20 meters |
| | Ⓐ  Ⓑ  Ⓒ  Ⓓ | |

11. **Multiple Choice** What are the zeros of the function $y = x^2 - 100$?
    - Ⓐ 10, 10
    - Ⓑ 100, 0
    - Ⓒ −10, 10
    - Ⓓ −10, −10

Algebra 1         Standardized Test Practice 9.8    **65**

# SAT/ACT Chapter Test
## Chapter 9  Polynomials and Factoring

**TEST TAKING STRATEGY**  Look for math ideas that are presented in words.

1. **Multiple Choice**  What are the solutions to the equation $0 = x^2 + 12x + 20$?
   - Ⓐ $-10, 2$
   - Ⓑ $-10, -2$
   - Ⓒ $10, -2$
   - Ⓓ $10, 2$

2. **Multiple Choice**  What are the two factors of the polynomial $9x^2 - 6x - 8$?
   - Ⓐ $(3x + 2)(3x - 4)$
   - Ⓑ $(9x - 2)(x + 4)$
   - Ⓒ $(3x - 2)(3x + 4)$
   - Ⓓ $(3x - 2)^2$

3. **Multiple Choice**  Which polynomial can *not* be factored?
   - Ⓐ $3x - 6$
   - Ⓑ $8x - 4$
   - Ⓒ $x^2 - 16$
   - Ⓓ $x^2 + 16$

4. **Multiple Choice**  Which sum is equal to $3x^3 - 6x^2 + 2x - 1$?
   - Ⓐ $(2x^3 - 4x^2 + x) + (x^3 - 2x^2 + x - 1)$
   - Ⓑ $(x^3 + 4x^2 + x + 3) + (2x^3 - 2x^2 + x - 1)$
   - Ⓒ $(-2x^3 - 4x^2 + x + 1) + (4x^3 - 2x^2 + x)$
   - Ⓓ $(3x^3 - 5x^2 + 2x) + (4x^3 - 5x^2 + 2x - 1)$

5. **Multiple Choice**  The area of a square is $x^2 + 16x + 64$. What is the perimeter?
   - Ⓐ $2x + 16$
   - Ⓑ $4x$
   - Ⓒ $x + 8$
   - Ⓓ $4x + 32$

6. **Multiple Choice**  Which polynomial function is true for $x$-values $-1, 0$ and $1$?
   - Ⓐ $x^2 + 5x + 6 = (x + 3)(x + 2)$
   - Ⓑ $x^2 + 64 = (x + 8)(x + 8)$
   - Ⓒ $x^2 - 14x + 28 = (x + 7)(x + 4)$
   - Ⓓ $x^2 + 10x + 50 = (x + 5)(x + 10)$

7. **Multiple Choice**  Which is *not* a possible value for $b$ in the trinomial $x^2 + bx - 25$?
   - Ⓐ $-24$
   - Ⓑ $24$
   - Ⓒ $0$
   - Ⓓ $25$

*Quantitative Comparison*  In Exercises 8–11, choose the letter of the statement below that is true about the quantities in Columns I and II.

**A** The number in Column I is greater.
**B** The number in Column II is greater.
**C** The two numbers are equal.
**D** The relationship cannot be determined from the given information.

| | Column I | Column II |
|---|---|---|
| 8. | the number of terms in a cubic trinomial | the number of terms in a linear polynomial |
| | Ⓐ  Ⓑ | Ⓒ  Ⓓ |
| 9. | the last term in the product $(x - 4)(x + 4)$ | 16 |
| | Ⓐ  Ⓑ | Ⓒ  Ⓓ |
| 10. | the number of factors of 24 | the number of factors of 30 |
| | Ⓐ  Ⓑ | Ⓒ  Ⓓ |
| 11. | the number of square units in the surface area of a cube with 6-centimeter sides | the number of cubic units in the volume of a cube with 6-centimeter sides |
| | Ⓐ  Ⓑ | Ⓒ  Ⓓ |

NAME _____ CLASS _____ DATE _____

# Standardized Test Practice
## 10.1 Graphing parabolas

**TEST TAKING STRATEGY** Eliminate impossible answers before guessing.

1. **Multiple Choice** In the quadratic function $y = ax^2 + bx + c$, $a$ cannot equal zero because:
   - Ⓐ $b$ or $c$ would then equal zero.
   - Ⓑ it would not have a reciprocal.
   - Ⓒ the equation would no longer be quadratic, it would be linear.
   - Ⓓ none of the above

2. **Multiple Choice** A parabola represented by $y = ax^2 + bx + c$, where $a < 0$, opens
   - Ⓐ upward.
   - Ⓑ to the right.
   - Ⓒ downward.
   - Ⓓ to the left.

3. **Multiple Choice** The function $y = x^2 + 2$ is what type of transformation of the parent function $y = x^2$?
   - Ⓐ vertical translation
   - Ⓑ horizontal translation
   - Ⓒ vertical stretch
   - Ⓓ horizontal compression

4. **Multiple Choice** The coordinates of the vertex of the graph of $y = (x + 1)^2 - 2$ are:
   - Ⓐ (1, 2)
   - Ⓑ (−1, 2)
   - Ⓒ (1, −2)
   - Ⓓ (−1, −2)

5. **Multiple Choice** The axis of symmetry of the graph of $y = \frac{1}{2}(x - 4)^2 + 3$ is:
   - Ⓐ $x = 4$
   - Ⓑ $x = -4$
   - Ⓒ $x = 2$
   - Ⓓ $x = -2$

6. **Multiple Choice** What are the zeros of the function $y = x^2 + 2x - 24$?
   - Ⓐ 6 and 4
   - Ⓑ −6 and 4
   - Ⓒ −6 and −4
   - Ⓓ 6 and −4

*Quantitative Comparison* In Exercises 7–10, choose the letter of the statement below that is true about the quantities in Columns I and II.

A The number in Column I is greater.
B The number in Column II is greater.
C The two numbers are equal.
D The relationship cannot be determined from the given information.

| Column I | Column II |
|---|---|
| 7. maximum height of the graph of a quadratic function | minimum height of the graph of a quadratic function |
| Ⓐ  Ⓑ | Ⓒ  Ⓓ |
| 8. $h = 40t - 5t^2$ | |
| the height, $h$, after $t = 3$ seconds | the height, $h$, after $t = 5$ seconds |
| Ⓐ  Ⓑ | Ⓒ  Ⓓ |
| 9. the height when something that is thrown hits the ground | the initial height when something is thrown into the air |
| Ⓐ  Ⓑ | Ⓒ  Ⓓ |
| 10. the distance of the vertex from the origin in the graph of $y = x^2 + 2$ | the distance of the vertex from the origin in the graph of $y = (x - 2)^2$ |
| Ⓐ  Ⓑ | Ⓒ  Ⓓ |

11. **Multiple Choice** Which describes the shape of the graph of a quadratic function?
   - Ⓐ circle
   - Ⓑ straight line
   - Ⓒ parabola
   - Ⓓ two parallel lines

NAME _____ CLASS _____ DATE _____

# Standardized Test Practice
## 10.2 Solving Equations by Using Square Roots

**TEST TAKING STRATEGY** Use number sense to eliminate unreasonable choices.

1. **Multiple Choice** Solve $x^2 = 10$ for $x$.
   - Ⓐ $x = 5$
   - Ⓑ $x = \sqrt{10}$
   - Ⓒ $x = \pm 5$
   - Ⓓ $x = \pm\sqrt{10}$

2. **Multiple Choice** Which value shows the solution to $(x + 4)^2 = 0$?
   - Ⓐ $x = 5$
   - Ⓑ $x = -4$
   - Ⓒ $x = 2$
   - Ⓓ $x = -2$

3. **Multiple Choice** The first step in solving the equation $(x - 2)^2 = 25$, is to:
   - Ⓐ take the square root of each side.
   - Ⓑ square the right side only.
   - Ⓒ subtract 2 from each side.
   - Ⓓ substitute 2 for $x$.

4. **Multiple Choice** Which equation has the solution $x = -10$ or $x = 10$?
   - Ⓐ $x^2 - 100 = 0$
   - Ⓑ $x - 100 = \sqrt{1000}$
   - Ⓒ $x + 100 = 10^2$
   - Ⓓ $10^2 + x^2 = 100^2$

5. **Multiple Choice** In the graph of $y^2 = (x - 3)^2 - 4$, the point $(3, -4)$ represents which of the following?
   - Ⓐ axis of symmetry
   - Ⓑ vertex
   - Ⓒ zeros
   - Ⓓ solution

6. **Multiple Choice** Solve $7(x + 6)^2 = 448$.
   - Ⓐ $x = 4$ or $x = -10$
   - Ⓑ $x = 12$ or $x = -6$
   - Ⓒ $x = 2$ or $x = -14$
   - Ⓓ $x = 10$ or $x = -10$

*Quantitative Comparison* In Exercises 7–10, choose the letter of the statement below that is true about the principal square root in Columns I and II.

- **A** The number in Column I is greater.
- **B** The number in Column II is greater.
- **C** The two numbers are equal.
- **D** The relationship cannot be determined from the given information.

| | Column I | Column II | |
|---|---|---|---|
| 7. | \multicolumn{2}{c}{the value of $x$ if $x > 0$} | |
| | $2x^2 = 56$ | $x^2 = 30$ | |
| | Ⓐ  Ⓑ | Ⓒ  Ⓓ | |
| 8. | \multicolumn{2}{c}{the value of $x$ if $x > 0$} | |
| | $-3x^2 = -35$ | $x^2 = \dfrac{121}{25}$ | |
| | Ⓐ  Ⓑ | Ⓒ  Ⓓ | |
| 9. | \multicolumn{2}{c}{the value of $x$ if $x < 0$} | |
| | $(x - 8)^2 = 81$ | $(x - 2)^2 = 144$ | |
| | Ⓐ  Ⓑ | Ⓒ  Ⓓ | |

10. **Multiple Choice** What is the axis of symmetry in the graph of $y = (x - 8)^2 - 81$?
    - Ⓐ $x = 8$
    - Ⓑ $x = -8$
    - Ⓒ $x = 81$
    - Ⓓ $x = -81$

11. **Multiple Choice** Find the time, $t$, in seconds that it takes for an object to fall $h = 500$ feet. Use the function $h = -16t^2 + 600$.
    - Ⓐ 10.4 seconds
    - Ⓑ 2.5 seconds
    - Ⓒ 100.1 seconds
    - Ⓓ 5.2 seconds

NAME _____ CLASS _____ DATE _____

# Standardized Test Practice
## 10.3 Completing the Square

**TEST TAKING STRATEGY** Use the test as an information tool to help answer other questions.

1. **Multiple Choice** Which diagram represents how to complete the square of $x^2 + 4x$ using algebra tiles?

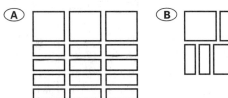

2. **Multiple Choice** Which equation shows $y = x^2 - 10x + 9$ in the form $y = (x - h)^2 + k$?
   - Ⓐ $y = (x - 3)^2$
   - Ⓑ $y = (x - 5)^2 + 4$
   - Ⓒ $y = (x + 3)^2$
   - Ⓓ $y = (x - 5)^2 - 16$

3. **Multiple Choice** If you complete the square for $x^2 + 16x$, the result is:
   - Ⓐ $(x - 8)^2$
   - Ⓑ $(x - 4)^2$
   - Ⓒ $(x + 8)^2$
   - Ⓓ $(x + 4)^2$

4. **Multiple Choice** Which point is the vertex in the graph of $y = x^2 - 3$?
   - Ⓐ $(0, 3)$
   - Ⓑ $(0, -3)$
   - Ⓒ $(3, 0)$
   - Ⓓ $(-3, 0)$

5. **Multiple Choice** Which value makes a perfect-square trinomial when added to the binomial $x^2 - 9x$?
   - Ⓐ $-3$
   - Ⓑ $3$
   - Ⓒ $\dfrac{9}{2}$
   - Ⓓ $\left(\dfrac{9}{2}\right)^2$

***Quantitative Comparison*** In Exercises 6–7, choose the letter of the statement below that is true about the quantities in Columns I and II.

**A** The number in Column I is greater.
**B** The number in Column II is greater.
**C** The two numbers are equal.
**D** The relationship cannot be determined from the given information.

| | Column I | | Column II | |
|---|---|---|---|---|
| 6. | the value of $c$ that makes each expression a perfect-square trinomial | | | |
| | $x^2 + 4x + c$ | | $x^2 - 8x + c$ | |
| | Ⓐ | Ⓑ | Ⓒ | Ⓓ |
| 7. | the value of $c$ that makes each expression a perfect-square trinomial | | | |
| | $x^2 - 9x + c$ | | $x^2 - 3x + c$ | |
| | Ⓐ | Ⓑ | Ⓒ | Ⓓ |

8. **Multiple Choice** Which shows $y = x^2 - 2x - 3$ written in vertex form?
   - Ⓐ $y = (x + 1)^2 + 4$
   - Ⓑ $y = (x + 1)^2 - 7$
   - Ⓒ $y = (x - 1)^2 - 2$
   - Ⓓ $y = (x - 1)^2 - 4$

9. **Multiple Choice** If you complete the square and rewrite the function $y = x^2 + 4x$ in vertex form, what is $k$?
   - Ⓐ $-4$
   - Ⓑ $-16$
   - Ⓒ $4$
   - Ⓓ $16$

10. **Multiple Choice** Find the vertex of the graph of $y = -x^2 + 2$.
    - Ⓐ $(-2, 0)$
    - Ⓑ $(0, 2)$
    - Ⓒ $(0, -2)$
    - Ⓓ $(2, 0)$

NAME _____ CLASS _____ DATE _____

# Standardized Test Practice
## 10.4 Solving Equations of the Form $x^2 + bx + c = 0$

**TEST TAKING STRATEGY**  Look over the test, begin with the questions you can do quickly.

1. **Multiple Choice**  To solve the equation $x^2 + 6x = 7$ by factoring, the first step is to:
   - Ⓐ take the square root of each side.
   - Ⓑ factor $x^2 + 6x$.
   - Ⓒ subtract 7 from each side.
   - Ⓓ subtract 6 from each side.

2. **Multiple Choice**  To solve the equation $x^2 + 3x + 2 = 0$ by completing the square, the first step involves:
   - Ⓐ taking the square root of each side.
   - Ⓑ factoring $x^2 + 3x$.
   - Ⓒ factoring $x^2 + 3x + 2$.
   - Ⓓ none of the above

3. **Multiple Choice**  What are the zeros of the function $y = x^2 + x - 2$?
   - Ⓐ $\frac{1}{2}$ and 0
   - Ⓑ $-1 \pm \sqrt{2}$
   - Ⓒ $-2$ and 1
   - Ⓓ $-1$ and 2

4. **Multiple Choice**  For the function $y = x^2 + 5x - 10$, which shows the solution for $x$ when $y = -4$?
   - Ⓐ $x = -5 + \sqrt{6}$ or $x = -5 - \sqrt{6}$
   - Ⓑ $x = -6$ or $x = 1$
   - Ⓒ $x = -1$ or $x = 6$
   - Ⓓ no solution

5. **Multiple Choice**  The graph of a function in the form of $y = ax^2 + bx + c$ can best be described as:
   - Ⓐ a circle.
   - Ⓑ a straight line.
   - Ⓒ a parabola.
   - Ⓓ two parallel lines.

**Quantitative Comparison**  In Exercises 6–8, choose the letter of the statement below that is true about the quantities in Columns I and II.

A  The number in Column I is greater.
B  The number in Column II is greater.
C  The two numbers are equal.
D  The relationship cannot be determined from the given information.

| Column I | Column II |
|---|---|
| 6. | $h = 40t - 5t^2 + 8$ |
| the height, $h$, after $t = 3$ seconds | the height, $h$, after $t = 5$ seconds |
| Ⓐ   Ⓑ | Ⓒ   Ⓓ |

7.  the number of zeros in the function

| $y = x^2 - 6x + 7$ | $y = x^2 + 2x + 5$ |
|---|---|
| Ⓐ   Ⓑ | Ⓒ   Ⓓ |

8.  the positive zero of the function:

| $y = x^2 - 2x - 3$ | $y = x^2 + 4x - 5$ |
|---|---|
| Ⓐ   Ⓑ | Ⓒ   Ⓓ |

9. **Multiple Choice**  Where do the graphs of $y = x^2 - 3x + 2$ and $y = x - 1$ intersect?
   - Ⓐ $(1, 0)$ and $(3, 2)$
   - Ⓑ $(-1, 0)$ and $(3, -2)$
   - Ⓒ $(1, 0)$ and $(-2, 3)$
   - Ⓓ $(-1, 0)$ and $(-2, -3)$

NAME _____ CLASS _____ DATE _____

# Standardized Test Practice
## 10.5 The Quadratic Formula

**TEST TAKING STRATEGY** Write down known formulas in the margin before you begin.

1. **Multiple Choice** Which of the following is the quadratic formula?
   - Ⓐ $ax^2 + bx + c = 0$
   - Ⓑ $y = mx + b$
   - Ⓒ $y - y_1 = m(x - x_1)$
   - Ⓓ $x = \dfrac{-b \pm \sqrt{b^2 - 4ac}}{2a}$

2. **Multiple Choice** Solve the equation $x^2 - 7 + 2x = 0$ by using the quadratic formula.
   - Ⓐ $x = -1 + 2\sqrt{2}$ or $x = -1 - 2\sqrt{2}$
   - Ⓑ $x = 1 + 2\sqrt{2}$ or $x = 1 - 2\sqrt{2}$
   - Ⓒ $x = \dfrac{-7 + \sqrt{41}}{2}$ or $x = \dfrac{-7 - \sqrt{41}}{2}$
   - Ⓓ $x = \dfrac{7 + \sqrt{41}}{2}$ or $x = \dfrac{7 - \sqrt{41}}{2}$

3. **Multiple Choice** Which method is not used for solving a quadratic equation?
   - Ⓐ taking the square root of each side
   - Ⓑ factoring
   - Ⓒ point-slope formula
   - Ⓓ quadratic formula

4. **Multiple Choice** The discriminant is defined as:
   - Ⓐ $ax^2 + bx + c = 0$
   - Ⓑ $b^2 - 4ac$
   - Ⓒ $x = \dfrac{-b \pm \sqrt{b^2 - 4ac}}{2a}$
   - Ⓓ none of the above

**Quantitative Comparison** In Exercises 5–7, choose the letter of the statement below that is true about the quantities in Columns I and II.

- **A** The number in Column I is greater.
- **B** The number in Column II is greater.
- **C** The two numbers are equal.
- **D** The relationship cannot be determined from the given information.

| | Column I | Column II |
|---|---|---|
| 5. | the number of real number solutions in the equation $3x^2 + 6x + 3 = 0$ | $-2x^2 + 7x = 5$ |
| | Ⓐ  Ⓑ | Ⓒ  Ⓓ |
| 6. | the number of zeros in the function $y = x^2 + 7x - 7$ | $y = 2x^2 - 8x + 10$ |
| | Ⓐ  Ⓑ | Ⓒ  Ⓓ |
| 7. | the value of the discriminant $x^2 + 2x - 8 = 0$ | $8x^2 - 2 = 0$ |
| | Ⓐ  Ⓑ | Ⓒ  Ⓓ |

8. **Multiple Choice** Which is the solution for $x$ in $x^2 - 13 = 0$?
   - Ⓐ $x = 13$ or $x = -13$
   - Ⓑ $x = 3.61$ or $x = -3.61$
   - Ⓒ $x = 16.5$ or $x = -16.5$
   - Ⓓ $x = 169$ or $x = -169$

9. **Multiple Choice** If the value of the discriminant is greater than zero, how many solutions exist?
   - Ⓐ 2
   - Ⓑ 1
   - Ⓒ 0
   - Ⓓ More than 2

Algebra 1

NAME _____ CLASS _____ DATE _____

# Standardized Test Practice
## 10.6 Graphing Quadratic Inequalities

**TEST TAKING STRATEGY** Read each question carefully, know what each question is asking.

1. **Multiple Choice** Which of the following is the solution to the inequality $x^2 + 10x + 25 > 0$?
   - Ⓐ $x > -5$
   - Ⓑ $x > -5$ or $x < -5$
   - Ⓒ $x < -5$
   - Ⓓ $-5 < x < 5$

2. **Multiple Choice** Which of the following values for $x$ are included in the solution of $x^2 - 9 < 0$?
   - Ⓐ $x = 9$
   - Ⓑ $x = -3$
   - Ⓒ $x = -8$
   - Ⓓ $x = 0$

3. **Multiple Choice** Which graph represents the inequality $y > x^2$?

   Ⓐ 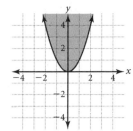   Ⓑ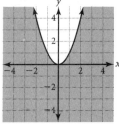

   Ⓒ 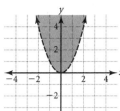   Ⓓ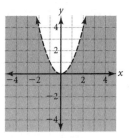

4. **Multiple Choice** Choose the best description of the graph of $y \leq x^2 - 3$.
   - Ⓐ dashed curve, shaded inside
   - Ⓑ solid curve, shaded inside
   - Ⓒ dashed curve, shaded outside
   - Ⓓ solid curve, shaded outside

**Quantitative Comparison** In Exercise 5, choose the letter of the statement below that is true about the quantities in Columns I and II.

A The number in Column I is greater.
B The number in Column II is greater.
C The two numbers are equal.
D The relationship cannot be determined from the given information.

| | Column I | Column II |
|---|---|---|
| 5. | the number of solutions in the inequality $x^2 - 8x - 20 < 0$ | the number of solutions in the inequality $2x^2 + 3x + 2 \leq 0$ |
| | Ⓐ    Ⓑ | Ⓒ    Ⓓ |

6. **Multiple Choice** The boundary line for a quadratic inequality is solid if the inequality is:

   I. $>$    II. $\geq$    III. $<$    IV. $\geq$

   - Ⓐ I and III
   - Ⓑ I only
   - Ⓒ II and IV
   - Ⓓ II only

7. **Multiple Choice** Choose the inequality that best describes the following graph:

   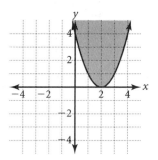

   - Ⓐ $y < (x - 2)^2$
   - Ⓑ $y < (x + 2)^2$
   - Ⓒ $y < x^2 - 2$
   - Ⓓ $y < x^2 + 2$

# SAT/ACT Chapter Test

## Chapter 10  Quadratic Functions

**TEST TAKING STRATEGY**  Use context to help define an unknown word.

1. **Multiple Choice** Which of the following shows the vertex and axis of symmetry for the graph of $y = (x + 3)^2 - 5$?

   Ⓐ $(-3, -5); x = -3$
   Ⓑ $(3, -5); x = 3$
   Ⓒ $(-3, 5); x = -3$
   Ⓓ $(3, 5); x = 3$

2. **Multiple Choice** The zeros of the function $y = x^2 - x - 6$ are:

   Ⓐ $-2$ and $-3$
   Ⓑ $2$ and $3$
   Ⓒ $-2$ and $3$
   Ⓓ $2$ and $-3$

3. **Multiple Choice** The function $y = 2x^2 - 8x$ has zeros at which of the following points?

   Ⓐ $(4, 0)$
   Ⓑ $(0, 0)$ and $(4, 0)$
   Ⓒ $(-4, 0)$
   Ⓓ $(0, 0)$ and $(-4, 0)$

4. **Multiple Choice** Which shows the function $y = x^2 - 2x + 5$ in vertex form?

   Ⓐ $y = (x - 1)^2 + 4$
   Ⓑ $y = (x - 1)^2 - 4$
   Ⓒ $y = (x + 1)^2 + 4$
   Ⓓ $y = (x + 1)^2 - 4$

5. **Multiple Choice** What is the value of the discriminant in $5x^2 - 6x + 2 = 0$?

   Ⓐ $4$
   Ⓑ $-4$
   Ⓒ $2$
   Ⓓ $-2$

**Quantitative Comparison** In Exercises 6–8, choose the letter of the statement below that is true about the quantities in Columns I and II.

A  The number in Column I is greater.
B  The number in Column II is greater.
C  The two numbers are equal.
D  The relationship cannot be determined from the given information.

| | Column I | Column II |
|---|---|---|
| 6. | the value of $c$ that makes a perfect square trinomial | |
| | $x^2 - 10x + c$ | $x^2 + 8x + c$ |
| | Ⓐ  Ⓑ | Ⓒ  Ⓓ |
| 7. | the number of real-number solutions | |
| | $4x^2 - 4x + 1 = 0$ | $x^2 + 6x - 12 = 0$ |
| | Ⓐ  Ⓑ | Ⓒ  Ⓓ |
| 8. | the positive solution of the equation | |
| | $x^2 - 4 = 0$ | $x^2 = 12$ |
| | Ⓐ  Ⓑ | Ⓒ  Ⓓ |

9. **Multiple Choice** Which is the solution of $x^2 + 2x = 24$?

   Ⓐ $x = 22$ or $x = 24$
   Ⓑ $x = -6$ or $x = 4$
   Ⓒ $x = -22$ or $x = -24$
   Ⓓ $x = 6$ or $x = -4$

10. **Multiple Choice** There are real number solutions of $x^2 + 2x + 35 = 0$.

    Ⓐ $2$
    Ⓑ $1$
    Ⓒ $0$
    Ⓓ more than 2

Algebra 1

# Standardized Test Practice
## 11.1 Inverse Variation

**TEST TAKING STRATEGY** Read each question carefully.

1. **Multiple Choice** The variation represented by "$y$ varies inversely as $x$" can be written as:
   - Ⓐ $y = kx$
   - Ⓑ $x = ky$
   - Ⓒ $y = \dfrac{k}{x}$
   - Ⓓ $x = \dfrac{y}{k}$

2. **Multiple Choice** If $y$ is 3 when $x$ is 6 and $y$ varies inversely as $x$ when $x$ is 2, what is the first step to be done in solving for $y$?
   - Ⓐ Solve for $y$.
   - Ⓑ Solve for $x$.
   - Ⓒ Solve for $k$.
   - Ⓓ none of the above

3. **Multiple Choice** If $y$ varies inversely as $x$, which equation shows how $y$ is related to $x$ if $y$ is 7 when $x$ is 50?
   - Ⓐ $y = \dfrac{50}{7}$
   - Ⓑ $y = \dfrac{x}{350}$
   - Ⓒ $y = \dfrac{7}{50}$
   - Ⓓ $y = \dfrac{350}{x}$

4. **Multiple Choice** Which equation shows the relationship between the variables?

   | $m$ | 36 | 18 | 9 | 6 |
   |---|---|---|---|---|
   | $n$ | 1 | 2 | 4 | 6 |

   - Ⓐ $m = 36n$
   - Ⓑ $n = \dfrac{36}{m}$
   - Ⓒ $m \cdot n = 36$
   - Ⓓ $36m = n$

5. **Multiple Choice** Which of the following equations represents an inverse variation?
   I. $mn = 300$  II. $\dfrac{x}{y} = \dfrac{1}{2}$  III. $\dfrac{x}{3} = \dfrac{4}{y}$
   - Ⓐ I only
   - Ⓑ II only
   - Ⓒ I and III
   - Ⓓ II and III

**Quantitative Comparison** In Exercises 6–7, choose the letter of the statement below that is true about the quantities in Columns I and II.

A The number in Column I is greater.
B The number in Column II is greater.
C The two numbers are equal.
D The relationship cannot be determined from the given information.

|  | Column I | Column II |
|---|---|---|
| 6. | \multicolumn{2}{c}{$y$ varies inversely as $x$} |
|  | the value of $x$ when $y$ is 4, if $y$ is 8 when $x$ is 2 | the value of $y$ when $x$ is $-6$, if $y$ is 3 when $x$ is $-8$ |
|  | Ⓐ  Ⓑ | Ⓒ  Ⓓ |
| 7. | \multicolumn{2}{c}{the constant, $k$} |
|  | $-8 = \dfrac{k}{3}$ | $7 = \dfrac{k}{-4}$ |
|  | Ⓐ  Ⓑ | Ⓒ  Ⓓ |

8. **Multiple Choice** Use the rule of 72 to estimate how long it would take to double a given amount of money at 9% interest, compounded annually.
   - Ⓐ 648 years
   - Ⓑ 8 years
   - Ⓒ 9 years
   - Ⓓ 81 years

9. **Multiple Choice** The frequency of a vibrating string is inversely proportional to its length. A violin string 10 inches long vibrates at a frequency of 512 cycles per second. Find the frequency of an 8-inch string.
   - Ⓐ 409.6 cycles per second
   - Ⓑ 640 cycles per second
   - Ⓒ 6.4 cycles per second
   - Ⓓ 40.96 cycles per second

NAME _____ CLASS _____ DATE _____

# Standardized Test Practice
## 11.2 Rational Expressions and Functions

**TEST TAKING STRATEGY** Treat each multiple choice question as a true/false question.

1. **Multiple Choice** Which choice best describes a rational expression?
   - Ⓐ fractions
   - Ⓑ repeating decimals
   - Ⓒ terminating decimals
   - Ⓓ all of the above

2. **Multiple Choice** What is the domain of the function $y = \frac{3}{x}$?
   - Ⓐ all real numbers
   - Ⓑ all real numbers except for 3
   - Ⓒ all real numbers except for 0
   - Ⓓ all real numbers except for $-3$

3. **Multiple Choice** Which of the following is an example of a non-trivial rational function?
   - Ⓐ $y = \frac{x}{2}$
   - Ⓑ $y = \frac{2}{x}$
   - Ⓒ $y = 5x$
   - Ⓓ $y = \frac{2}{3}x$

4. **Multiple Choice** Which answer represents the function $y = \frac{6x + 9x^2}{x^2}$ written in simplest form?
   - Ⓐ $y = \frac{6}{x} + 9x$
   - Ⓑ $y = \frac{6}{x} + 9$
   - Ⓒ $y = 6x + 9$
   - Ⓓ $y = \frac{6}{x} + 9x^2$

**Quantitative Comparison** In Exercises 5–7, choose the letter of the statement below that is true about the quantities in Columns I and II.

- **A** The number in Column I is greater.
- **B** The number in Column II is greater.
- **C** The two numbers are equal.
- **D** The relationship cannot be determined from the given information.

| | Column I | Column II |
|---|---|---|
| 5. | the value of $x$ such that the expression is undefined | |
| | $\dfrac{3x + 9}{x}$ | $\dfrac{4x - 5}{3x + 6}$ |
| | Ⓐ    Ⓑ | Ⓒ    Ⓓ |
| 6. | the value of the variable such that the expression is undefined | |
| | $\dfrac{x^2 - 6x + 8}{x - 2}$ | $\dfrac{2w^2 + 4w}{2w + 5}$ |
| | Ⓐ    Ⓑ | Ⓒ    Ⓓ |
| 7. | the value of the function $y = \dfrac{3x}{x - 1}$ when $x = 4$ | the value of the function $y = \dfrac{2}{3x - 4} + 7x$ when $x = 2$ |
| | Ⓐ    Ⓑ | Ⓒ    Ⓓ |

8. **Multiple Choice** Which answer represents the function $y = \dfrac{x^{50} + x^{48}}{x^{48}}$ written in simplest form?
   - Ⓐ $x^2$
   - Ⓑ $x^2 - x$
   - Ⓒ $x^2 + 1$
   - Ⓓ $\dfrac{1}{x^{46}}$

Algebra 1  Standardized Test Practice 11.2

NAME _____ CLASS _____ DATE _____

# Standardized Test Practice
## 11.3 Simplifying Rational Expressions

**TEST TAKING STRATEGY** Use number sense to eliminate unreasonable choices.

1. **Multiple Choice** What are the restrictions on $x$ in $\dfrac{6x - 12}{3}$?

   Ⓐ $x \neq 0$   Ⓑ $x \neq 2$
   Ⓒ $x \neq -2$   Ⓓ no restrictions

2. **Multiple Choice** What are the restrictions on $y$ in $\dfrac{y^2 - 36}{y^2 + 3y - 18}$?

   Ⓐ $y \neq 3$
   Ⓑ $y \neq -6$ and $y \neq 3$
   Ⓒ $y \neq -6$
   Ⓓ $y \neq 6$ and $y \neq -6$

3. **Multiple Choice** Which represents the expression $\dfrac{x^2 - x - 56}{x^2 + x - 42}$ written in simplest form?

   Ⓐ $\dfrac{4}{3}$   Ⓑ $-x - 14$
   Ⓒ $\dfrac{x - 8}{x - 6}$   Ⓓ $\dfrac{x + 8}{6}$

4. **Multiple Choice** Which is equivalent to $\dfrac{-a - 1}{a^2 + 7a + 6}$?

   Ⓐ $\dfrac{-1}{a + 6}$   Ⓑ $\dfrac{-1}{a + 3}$
   Ⓒ $\dfrac{-1}{a + 2}$   Ⓓ $\dfrac{1}{a - 6}$

5. **Multiple Choice** What should you do first when simplifying a rational expression?

   Ⓐ Reduce the numbers.
   Ⓑ Reduce the variables.
   Ⓒ Factor the numerator and denominator.
   Ⓓ State any restrictions.

**Quantitative Comparison** In Exercises 6–8, choose the letter of the statement below that is true about the quantities in Columns I and II.

A The number in Column I is greater.
B The number in Column II is greater.
C The two numbers are equal.
D The relationship cannot be determined from the given information.

| Column I | Column II |
|---|---|

6. the greatest common factor of the numerator and denominator in the expression

   $\dfrac{9}{21}$  |  $\dfrac{6x + 9}{12}$

   Ⓐ   Ⓑ   Ⓒ   Ⓓ

7. the number of restrictions on the variable in the expression

   $\dfrac{x - 5}{x^2 - 5x}$  |  $\dfrac{m^2 - 4m}{m^2 - 6m + 9}$

   Ⓐ   Ⓑ   Ⓒ   Ⓓ

8. the values of the variables such that the expression is undefined

   $\dfrac{7p}{p - 4}$  |  $\dfrac{3(a + b)}{6(a - b)}$

   Ⓐ   Ⓑ   Ⓒ   Ⓓ

9. **Multiple Choice** Which expression is not in simplest form?

   Ⓐ $\dfrac{x + 3}{x}$   Ⓑ $\dfrac{x + 6}{3}$
   Ⓒ $\dfrac{2x(x - 5)}{4x^2}$   Ⓓ $\dfrac{3x - 6}{x - 2}$

NAME _____ CLASS _____ DATE _____

# Standardized Test Practice
## 11.4 Operations With Rational Expressions

**TEST TAKING STRATEGY** Be careful of distracters. They are usually included in the choices.

1. **Multiple Choice** Which of the following represents the restrictions on the variable in $\dfrac{x-2}{3x} \cdot \dfrac{x^2+3x}{x+3} \cdot \dfrac{4x}{5}$?

   Ⓐ $x \neq 0$
   Ⓑ $x \neq 0$ and $x \neq 3$
   Ⓒ $x \neq -3$
   Ⓓ $x \neq 0$ and $x \neq -3$

2. **Multiple Choice** The restrictions on the variable in $\dfrac{x-6}{5} \div \dfrac{5x+10}{x+3}$ are defined as:

   Ⓐ $x \neq -3$
   Ⓑ $x \neq 0$ and $x \neq -3$
   Ⓒ $x \neq 6$
   Ⓓ $x \neq -2$ and $x \neq 0$

3. **Multiple Choice** In the expression $\dfrac{5}{x^2-5x} + \dfrac{x}{x-2}$, $x$ can *not* be equal to:

   Ⓐ 0 and 2
   Ⓑ 0 and 5
   Ⓒ 2 and 5
   Ⓓ 0, 2, and 5

4. **Multiple Choice** Which expression represents $\dfrac{h^2-81}{h^2-36} \div \dfrac{h-9}{h+6}$ written in simplest form?

   Ⓐ $\dfrac{h+9}{h-6}$
   Ⓑ $\dfrac{h-6}{h+9}$
   Ⓒ $\dfrac{h+9}{h+6}$
   Ⓓ $\dfrac{3}{2}$

**Quantitative Comparison** In Exercises 5–6, choose the letter of the statement below that is true about the quantities in Columns I and II.

A The number in Column I is greater.
B The number in Column II is greater.
C The two numbers are equal.
D The relationship cannot be determined from the given information.

| | Column I | Column II |
|---|---|---|
| 5. | the number of restrictions on the variable in the expression $\dfrac{2a^2}{a^2+a} \cdot \dfrac{a+1}{a^3}$ | $\dfrac{x+5}{3x+6} - \dfrac{x-3}{x-2}$ |

   Ⓐ   Ⓑ   Ⓒ   Ⓓ

| | Column I | Column II |
|---|---|---|
| 6. | the number of restrictions on the variable in the expression $\dfrac{x+2}{x(x+1)} \div \dfrac{x+3}{x^2}$ | $\dfrac{-3}{c-5} + \dfrac{-7}{c^2-5c}$ |

   Ⓐ   Ⓑ   Ⓒ   Ⓓ

7. **Multiple Choice** Which of the following expressions is not equivalent to the other three?

   Ⓐ $\dfrac{x}{4} + \dfrac{x-y}{2}$
   Ⓑ $\dfrac{x^2-y^2}{8} \div \dfrac{x+y}{4}$
   Ⓒ $\dfrac{3}{x-y} \cdot \dfrac{(x-y)^2}{6}$
   Ⓓ $\dfrac{x}{2} - \dfrac{2y}{4}$

Algebra 1                    Standardized Test Practice 11.4        77

NAME _____ CLASS _____ DATE _____

# Standardized Test Practice
## 11.5 Solving Rational Equations

**TEST TAKING STRATEGY** Eliminate impossible answers before guessing.

1. **Multiple Choice** Which of the following is a possible method for solving rational equations?
   - Ⓐ graphing method
   - Ⓑ common denominator method
   - Ⓒ make a table
   - Ⓓ all of the above

2. **Multiple Choice** Which of the following represents the least common denominator for the equation $\dfrac{p^2}{p-5} + \dfrac{5}{5-p} = 2$?
   - Ⓐ $2(p-5)$
   - Ⓑ $(p-5)(p+5)$
   - Ⓒ $p-5$
   - Ⓓ $(p-5)(5-p)$

3. **Multiple Choice** For what value(s) of the variable is the equation $\dfrac{5}{x} + \dfrac{4}{x-5} = \dfrac{1}{5}$ undefined?
   - Ⓐ $x \neq 0$ and $x \neq -5$
   - Ⓑ $x \neq 0$ and $x \neq 5$
   - Ⓒ $x \neq 5$
   - Ⓓ $x \neq 0$ and $x \neq -5$ and $x \neq 5$

4. **Multiple Choice** Clark and Lance can clean the garage together in $3\tfrac{3}{5}$ hours. Clark can do the job alone in 6 hours. How many hours would it take Lance to do the job alone?
   - Ⓐ $1\tfrac{1}{5}$ hours
   - Ⓑ $2\tfrac{2}{5}$ hours
   - Ⓒ 9 hours
   - Ⓓ 2 hours

*Quantitative Comparison* In Exercises 5–6, choose the letter of the statement below that is true about the quantities in Columns I and II.

- **A** The number in Column I is greater.
- **B** The number in Column II is greater.
- **C** The two numbers are equal.
- **D** The relationship cannot be determined from the given information.

| | Column I | Column II |
|---|---|---|
| 5. | the least common denominator in the equation $\dfrac{w+3}{w} + \dfrac{4}{5w} = \dfrac{6}{5}$ | $\dfrac{4}{w-5} + \dfrac{3}{w} = \dfrac{4}{5}$ |
| | Ⓐ   Ⓑ | Ⓒ   Ⓓ |
| 6. | the number of solutions to the equation $\dfrac{6}{v-1} + 2 = \dfrac{12}{v^2-1}$ | $\dfrac{x+2}{x-2} - \dfrac{2}{x+2} = \dfrac{-7}{3}$ |
| | Ⓐ   Ⓑ | Ⓒ   Ⓓ |

7. **Multiple Choice** What is the solution to the rational equation $\dfrac{y-1}{y+1} - \dfrac{2y}{y-1} = -1$?
   - Ⓐ $x = 1$
   - Ⓑ $x = 0$
   - Ⓒ $x = -1$
   - Ⓓ no solution

8. **Multiple Choice** Solve for $x$.
   $$\dfrac{1}{x} + \dfrac{3}{x} = \dfrac{5}{4}$$
   - Ⓐ $x = 2$
   - Ⓑ $x = 3\tfrac{1}{5}$
   - Ⓒ $x = 4.5$
   - Ⓓ $x = 16$

NAME _____ CLASS _____ DATE _____

# Standardized Test Practice
## 11.6 Proof In Algebra

**TEST TAKING STRATEGY** Use context to define unknown words.

1. **Multiple Choice** In the conditional statement "if $p$, then $q$", what does $p$ represent?
   - Ⓐ a theorem
   - Ⓑ a conclusion
   - Ⓒ a proof
   - Ⓓ a hypothesis

2. **Multiple Choice** What is represented by the statement "if $q$, then $p$?"
   - Ⓐ a proof
   - Ⓑ inductive reasoning
   - Ⓒ a converse
   - Ⓓ deductive reasoning

3. **Multiple Choice** What can also be described as "an educated guess"?
   - Ⓐ a conjecture
   - Ⓑ deductive reasoning
   - Ⓒ a proof
   - Ⓓ a theorem

4. **Multiple Choice** Which of the following would be the reason for the following statement?
   $$2a + 0 = 2a$$
   - Ⓐ property of closure
   - Ⓑ additive identity
   - Ⓒ additive inverse
   - Ⓓ multiplicative identity

5. **Multiple Choice** Which of the following is an example of the Commutative Property of Addition?
   - Ⓐ $3m + 0 = 0 + 4$
   - Ⓑ $6(a + b) = (a + b)6$
   - Ⓒ $7 + 6m = 6m + 7$
   - Ⓓ $5(a + 2c) = 5a + 10c$

*Quantitative Comparison* In Exercises 6–7, choose the letter of the statement below that is true about the quantities in Columns I and II.

A The number in Column I is greater.
B The number in Column II is greater.
C The two numbers are equal.
D The relationship cannot be determined from the given information.

| | Column I | Column II |
|---|---|---|
| 6. | $2n$ | $2n + 1$ |
| | Ⓐ Ⓑ | Ⓒ Ⓓ |
| 7. | the number of operations that hold true for the Associative Property | the number of operations that hold true for the Closure Property |
| | Ⓐ Ⓑ | Ⓒ Ⓓ |

8. **Multiple Choice** Which of the following is *not* closed under multiplication?
   - Ⓐ $\{-2, 0, 2\}$
   - Ⓑ even numbers
   - Ⓒ integers
   - Ⓓ whole numbers

9. **Multiple Choice** For how many operations must a set maintain properties in order to be considered a field?
   - Ⓐ 1
   - Ⓑ 2
   - Ⓒ 3
   - Ⓓ 4

10. **Multiple Choice** Which of the following is *not* a type of reason that can be used in a proof?
    - Ⓐ conjectures
    - Ⓑ definitions
    - Ⓒ axioms
    - Ⓓ postulates

Algebra 1     Standardized Test Practice 11.6     79

# SAT/ACT Chapter Test
## Chapter 11  Rational Functions

**TEST TAKING STRATEGY**  Be sure you understand what is being asked.

1. **Multiple Choice**  If $y$ varies inversely as $x$ and $y = -6$ when $x = -2$, what is $y$ when $x = 9$?

   Ⓐ $x = \dfrac{3}{4}$   Ⓑ $x = \dfrac{4}{3}$
   Ⓒ $x = -27$   Ⓓ $x = 27$

2. **Multiple Choice**  What is the value of the expression $\dfrac{-4}{x^2 - x - 6}$ when $x = -2$?

   Ⓐ $-4$   Ⓑ $4$
   Ⓒ $\dfrac{2}{5}$   Ⓓ undefined

3. **Multiple Choice**  For what values of the variable is the function $y = \dfrac{-2}{x - 3} + x$ undefined?

   Ⓐ $x = 3$   Ⓑ $x = 0$ or $x = 3$
   Ⓒ $x = -3$   Ⓓ $x = 0$ or $x = -3$

4. **Multiple Choice**  What are the restrictions on $x$ in the expression $\dfrac{2x + 6}{6x^2 - 30x}$?

   Ⓐ $x \neq 5$   Ⓑ $x \neq 0$ and $x \neq 5$
   Ⓒ $x \neq -5$   Ⓓ $x \neq 0$ and $x \neq -5$

5. **Multiple Choice**  Which is equivalent to the expression $\dfrac{1}{3} + \dfrac{10}{5x}$?

   Ⓐ $\dfrac{11}{8x}$   Ⓑ $\dfrac{x + 6}{3x}$
   Ⓒ $\dfrac{3}{x}$   Ⓓ $\dfrac{2}{x + 3}$

**Quantitative Comparison**  In Exercises 6–8, choose the letter of the statement below that is true about the quantities in Columns I and II.

A  The number in Column I is greater.
B  The number in Column II is greater.
C  The two numbers are equal.
D  The relationship cannot be determined from the given information.

| | Column I | Column II |
|---|---|---|
| 6. | the value of $b$ | |
| | $\dfrac{2b - 3}{7} - \dfrac{b}{2} = \dfrac{b + 3}{14}$ | $\dfrac{2b}{3b + 6} - \dfrac{b}{5b + 10} = -\dfrac{2}{5}$ |
| | Ⓐ   Ⓑ | Ⓒ   Ⓓ |
| 7. | the number of restrictions on the variable in the expression | |
| | $\dfrac{x + y}{5x} \cdot \dfrac{3x}{x^2 - y^2}$ | $\dfrac{16}{y^2 - 16} - \dfrac{2}{y + 4}$ |
| | Ⓐ   Ⓑ | Ⓒ   Ⓓ |
| 8. | $y$ varies inversely as $x$ | |
| | the value of $x$ when $y$ is 4, if $y$ is 3 when $x$ is 12 | the value of $y$ when $x$ is 12, if $y$ is 16 when $x$ is 4 |
| | Ⓐ   Ⓑ | Ⓒ   Ⓓ |

9. **Multiple Choice**  If the area $\left(\dfrac{1}{2} \cdot \text{base} \cdot \text{height}\right)$ of a triangle is given by $x^2 - 4x - 21$ and its height is given by $x - 7$, what is the expression for the length of its base?

   Ⓐ $x - 3$   Ⓑ $x + 3$
   Ⓒ $2x - 6$   Ⓓ $2x + 6$

# Standardized Test Practice
## 12.1 Operations with Radicals

**TEST TAKING STRATEGY** Estimate to check that your answer is reasonable.

1. **Multiple Choice** What is $\sqrt{32}$ in simplest radical form?
   - Ⓐ $2\sqrt{8}$
   - Ⓑ $2\sqrt{2}$
   - Ⓒ $4\sqrt{2}$
   - Ⓓ $8\sqrt{2}$

2. **Multiple Choice** Which of the following is equivalent to $\sqrt{72m^2n^5}$?
   - Ⓐ $6mn\sqrt{2n^3}$
   - Ⓑ $6mn^2\sqrt{2n}$
   - Ⓒ $3mn\sqrt{8n^3}$
   - Ⓓ $3mn^2\sqrt{8n}$

3. **Multiple Choice** If you simplify the following expression, what is the result?
   $$\sqrt{3}(6 + \sqrt{8})$$
   - Ⓐ $6\sqrt{3} + \sqrt{24}$
   - Ⓑ $5\sqrt{6}$
   - Ⓒ $\sqrt{42}$
   - Ⓓ $6\sqrt{3} + 2\sqrt{6}$

4. **Multiple Choice** What is the product of $(4 + \sqrt{3})(1 - \sqrt{2})$?
   - Ⓐ $4 - \sqrt{6}$
   - Ⓑ $4 - 4\sqrt{2} + \sqrt{3} - \sqrt{6}$
   - Ⓒ $4 - 4\sqrt{5}$
   - Ⓓ $4 - 3\sqrt{5} - \sqrt{6}$

5. **Multiple Choice** The area of a small pond is equal to $\dfrac{9 + \sqrt{18}}{3}$ square meters. What other value is equivalent to this measure?
   - Ⓐ $3 + \sqrt{6}$ square meters
   - Ⓑ $3 + \sqrt{2}$ square meters
   - Ⓒ $3 + 3\sqrt{2}$ square meters
   - Ⓓ $9 + \sqrt{6}$ square meters

**Quantitative Comparison** In Exercises 6–9, choose the letter of the statement below that is true about the quantities in Columns I and II.

**A** The number in Column I is greater.
**B** The number in Column II is greater.
**C** The two numbers are equal.
**D** The relationship cannot be determined from the given information.

| | Column I | Column II |
|---|---|---|
| 6. | $\sqrt{\dfrac{4}{9}}$ | $\dfrac{\sqrt{50}}{\sqrt{8}}$ |
| | Ⓐ Ⓑ | Ⓒ Ⓓ |
| 7. | $4\sqrt{5} + 2\sqrt{5} - 5\sqrt{5}$ | $6\sqrt{3} - 4\sqrt{3} + 3\sqrt{3}$ |
| | Ⓐ Ⓑ | Ⓒ Ⓓ |
| 8. | $\sqrt{a^6b^8}$ | $\sqrt{x^{10}y^8}$ |
| | Ⓐ Ⓑ | Ⓒ Ⓓ |
| 9. | $(5\sqrt{2})^2$ | $(4\sqrt{3})^2$ |
| | Ⓐ Ⓑ | Ⓒ Ⓓ |

10. **Multiple Choice** What is the measure of the side of a square whose area is 360 square meters?
    - Ⓐ 60 meters
    - Ⓑ $6\sqrt{10}$ meters
    - Ⓒ 90 meters
    - Ⓓ $10\sqrt{36}$ meters

11. **Multiple Choice** Which expression is in simplest radical form?
    - Ⓐ $24 - \sqrt{36}$
    - Ⓑ $\sqrt{x^4} - 21$
    - Ⓒ $7 - 14\sqrt{5}$
    - Ⓓ $3\sqrt{500} - \sqrt{3}$

Algebra 1

NAME _____ CLASS _____ DATE _____

# Standardized Test Practice
## 12.2 Square-Root Functions and Radical Expressions

**TEST TAKING STRATEGY** Read the instructions to each question carefully.

1. **Multiple Choice** What is the solution to the equation $\sqrt{x+3} = 5$?
   - Ⓐ $x = 22$
   - Ⓑ $x = 2$
   - Ⓒ $x = 16$
   - Ⓓ $x = -4$

2. **Multiple Choice** Which equation is *not* possible to solve?
   - Ⓐ $\sqrt{x+12} = x$
   - Ⓑ $\sqrt{x+14} = 2$
   - Ⓒ $\sqrt{x-1} = x - 6$
   - Ⓓ $625 = 5x^2$

3. **Multiple Choice** Which values of $x$ make the equation $x^2 + 4x + 2 = 0$ true?
   - Ⓐ $x = -1$ or $x = -2$
   - Ⓑ $x = 2 \pm \sqrt{2}$
   - Ⓒ $x = -2 \pm \sqrt{2}$
   - Ⓓ $x = \dfrac{\sqrt{2}}{2}$

4. **Multiple Choice** Which of the following best describes the radius of a circular pool whose area is 25 square yards?
   - Ⓐ 2.82 yards
   - Ⓑ 5.00 yards
   - Ⓒ 1.59 yards
   - Ⓓ 8.86 yards

5. **Multiple Choice** Which ordered pair is on the graph of the function $y = -\sqrt{8-x}$?
   - Ⓐ $(4, -2)$
   - Ⓑ $(6, -2)$
   - Ⓒ $(-8, 4)$
   - Ⓓ $(8, -1)$

*Quantitative Comparison* In Exercises 6–8, choose the letter of the statement below that is true about the quantities in Columns I and II.

- **A** The number in Column I is greater.
- **B** The number in Column II is greater.
- **C** The two numbers are equal.
- **D** The relationship cannot be determined from the given information.

| | Column I | Column II |
|---|---|---|
| 6. | the number of solutions to the equation $x^2 = 85$ | the number of solutions to the equation $\sqrt{x+3} = x - 3$ |
| | Ⓐ   Ⓑ | Ⓒ   Ⓓ |
| 7. | the value of $c$ in the equation $16c^2 = 7$ | the value of $c$ in the equation $25c^2 = 3$ |
| | Ⓐ   Ⓑ | Ⓒ   Ⓓ |
| 8. | the solution to the equation $\sqrt{w} = 25$ | the solution to the equation $\sqrt{w} = -25$ |
| | Ⓐ   Ⓑ | Ⓒ   Ⓓ |

9. **Multiple Choice** Which of the following is the solution to the equation $|x| = 3$?
   - Ⓐ $x = 3$
   - Ⓑ $x = 3$ or $x = -3$
   - Ⓒ $x = -3$
   - Ⓓ no solution

10. **Multiple Choice** The solution to $x^2 = 3^2 + 5^2$ in simplified form is:
    - Ⓐ $x = 8$
    - Ⓑ $x = 6$
    - Ⓒ $x = \sqrt{34}$
    - Ⓓ $x = 2\sqrt{2}$

# Standardized Test Practice
## 12.3 The Pythagorean Theorem

**TEST TAKING STRATEGY** Write down known formulas before you begin.

**1. Multiple Choice** What is the missing side length in the following right triangle?

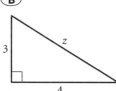

- Ⓐ 50 centimeters
- Ⓑ $5\sqrt{2}$ centimeters
- Ⓒ 25 centimeters
- Ⓓ $2\sqrt{5}$ centimeters

**2. Multiple Choice** For which triangle can you *not* use the Pythagorean Theorem?

Ⓐ    Ⓑ

Ⓒ    Ⓓ

**3. Multiple Choice** Which set of numbers represents the side lengths of a right triangle?

- Ⓐ 4, 5, 7
- Ⓑ 4, $4\sqrt{3}$, 8
- Ⓒ 3, 9, 7
- Ⓓ 5, $5\sqrt{2}$, 9

**Quantitative Comparison** In Exercises 4–5, choose the letter of the statement below that is true about the quantities in Columns I and II.

- **A** The number in Column I is greater.
- **B** The number in Column II is greater.
- **C** The two numbers are equal.
- **D** The relationship cannot be determined from the given information.

| | Column I | Column II |
|---|---|---|
| 4. | the length of the missing side of $\triangle ABC$ if m$\angle C = 90°$, $a = 3$ and $b = 9$ | the length of the missing side of $\triangle ABC$ if m$\angle B = 90°$, $a = 5$ and $b = 16$ |
| | Ⓐ    Ⓑ | Ⓒ    Ⓓ |
| 5. | the length of the hypotenuse of a right triangle whose shorter leg is 7 units | the length of the hypotenuse of a right triangle whose longer leg is 11 units |
| | Ⓐ    Ⓑ | Ⓒ    Ⓓ |

**6. Multiple Choice** Which set of numbers *cannot* represent the side lengths of a right triangle?

- Ⓐ 6, 8, 10
- Ⓑ 3, $3\sqrt{2}$, $3\sqrt{3}$
- Ⓒ 6, 9, 12
- Ⓓ $\sqrt{5}$, $\sqrt{7}$, $2\sqrt{3}$

**7. Multiple Choice** What is the measure of the altitude of an equilateral triangle whose side measures 10 inches?

- Ⓐ 5 inches
- Ⓑ $5\sqrt{3}$ inches
- Ⓒ 10 inches
- Ⓓ $5\sqrt{2}$ inches

NAME _____ CLASS _____ DATE _____

# Standardized Test Practice
## 12.4 The Distance Formula

**TEST TAKING STRATEGY** Use number sense to eliminate unreasonable choices.

1. **Multiple Choice** Which of the following is the distance formula?
   - Ⓐ $y - y_1 = m(x - x_1)$
   - Ⓑ $x = \dfrac{-b \pm \sqrt{b^2 - 4ac}}{2a}$
   - Ⓒ $d = \sqrt{(x_2 - x_1)^2 + (y_2 - y_1)^2}$
   - Ⓓ $\left(\dfrac{x_1 + x_2}{2}, \dfrac{y_1 + y_2}{2}\right)$

2. **Multiple Choice** What is the distance between the points $(3, 2)$ and $(6, -3)$?
   - Ⓐ $d = 64$
   - Ⓑ $d = \sqrt{34}$
   - Ⓒ $d = 8$
   - Ⓓ $d = \sqrt{10}$

3. **Multiple Choice** What is the ordered pair for the midpoint between the points $(6, 3)$ and $(-4, -5)$?
   - Ⓐ $(1, 1)$
   - Ⓑ $(2, -2)$
   - Ⓒ $(-5, -4)$
   - Ⓓ $(1, -1)$

4. **Multiple Choice** Which set of points represents the vertices of a right triangle?
   - Ⓐ $A(1, 2), B(3, 5), C(4, -1)$
   - Ⓑ $A(5, 0), B(0, 5), C(2, 0)$
   - Ⓒ $A(-4, 2), B(2, 4), C(2, 2)$
   - Ⓓ $A(-4, 0), B(0, 3), C(1, 1)$

5. **Multiple Choice** What is the midpoint of $\overline{PQ}$, given $P(7, -5)$ and $Q(-4, 8)$?
   - Ⓐ $\left(-\dfrac{11}{2}, -\dfrac{13}{2}\right)$
   - Ⓑ $\left(\dfrac{3}{2}, \dfrac{3}{2}\right)$
   - Ⓒ $\left(-\dfrac{3}{2}, \dfrac{13}{2}\right)$
   - Ⓓ $\left(\dfrac{11}{2}, \dfrac{3}{2}\right)$

**Quantitative Comparison** In Exercises 6–7, choose the letter of the statement below that is true about the quantities in Columns I and II.

- **A** The number in Column I is greater.
- **B** The number in Column II is greater.
- **C** The two numbers are equal.
- **D** The relationship cannot be determined from the given information.

| Column I | Column II |
|---|---|
| 6. the distance of $\overline{AB}$, given $A(2, 3)$ and $B(-4, -5)$ | the distance of $\overline{CD}$, given $C(5, 6)$ and $D(1, 2)$ |
| Ⓐ   Ⓑ | Ⓒ   Ⓓ |
| 7. the $x$-coordinate of the midpoint of $\overline{AB}$, given $A(7, -3)$ and $B(-4, 2)$ | the $y$-coordinate of the midpoint of $\overline{CD}$, given $C(4, -1)$ and $D(2, 4)$ |
| Ⓐ   Ⓑ | Ⓒ   Ⓓ |

8. **Multiple Choice** Given the midpoint $M(5, -2)$ and the endpoint $B(3, 6)$, what is the other endpoint of $\overline{AB}$?
   - Ⓐ $(7, -10)$
   - Ⓑ $(4, 2)$
   - Ⓒ $(2, -8)$
   - Ⓓ $(10, -4)$

9. **Multiple Choice** The midpoint of $\overline{PQ}$ is $M$. What are the missing coordinates? $P(-5, 2), Q(?, ?)$  $M(8, -8)$
   - Ⓐ $\left(\dfrac{3}{2}, -3\right)$
   - Ⓑ $\left(\dfrac{3}{2}, -4\right)$
   - Ⓒ $(11, -14)$
   - Ⓓ $(21, -18)$

NAME _____ CLASS _____ DATE _____

# Standardized Test Practice
## 12.5 Geometric Properties

**TEST TAKING STRATEGY** Look at each answer choice before selecting one.

1. *Multiple Choice* Which formula represents the equation of a circle?
   - (A) $y - y_1 = m(x - x_1)$
   - (B) $(x - h)^2 + (y - k)^2 = r^2$
   - (C) $d = \sqrt{(x_2 - x_1)^2 + (y_2 - y_1)^2}$
   - (D) $\left(\dfrac{x_1 + x_2}{2}, \dfrac{y_1 + y_2}{2}\right)$

2. *Multiple Choice* What is true about the segment joining the midpoints of two sides of a triangle?
   - (A) The segment is parallel to the third side.
   - (B) The segment is half the length of the third side.
   - (C) both A and B
   - (D) neither A nor B

3. *Multiple Choice* What is the equation of a circle whose center is $(3, -5)$ and whose radius is $\sqrt{5}$ inches?
   - (A) $(x - 3)^2 + (y + 5)^2 = 5$
   - (B) $(x - 3)^2 + (y - 5)^2 = 5$
   - (C) $(x - 3)^2 + (y - 5)^2 = 25$
   - (D) $(x + 3)^2 + (y - 5)^2 = 5$

4. *Multiple Choice* What is the center and radius of the circle represented by the following equation?
   $$(x - 7)^2 + (y + 3)^2 = \dfrac{4}{25}$$
   - (A) $(7, 3); r = 0.25$
   - (B) $(7, -3); r = 0.4$
   - (C) $(-7, 3); r = 0.4$
   - (D) $(7, -3); r = 0.25$

*Quantitative Comparison* In Exercises 5–6, choose the letter of the statement below that is true about the quantities in Columns I and II.

- **A** The number in Column I is greater.
- **B** The number in Column II is greater.
- **C** The two numbers are equal.
- **D** The relationship cannot be determined from the given information.

| | Column I | Column II |
|---|---|---|
| 5. | $r = 5\sqrt{2}$ | $r = 3\sqrt{7}$ |
|  | (A)  (B) | (C)  (D) |
| 6. | the measure of the radius of the circle represented by | |
|  | $x^2 + (y - 4)^2 = 35$ | $(x + 3)^2 + y^2 = 39.42$ |
|  | (A)  (B) | (C)  (D) |

7. *Multiple Choice* What is the center of the circle represented by the following equation?
   $$x^2 + y^2 = 25$$
   - (A) $(0, 5)$
   - (B) $(5, 0)$
   - (C) $(0, 0)$
   - (D) $(10, 10)$

8. *Multiple Choice* What is a directrix?
   - (A) a point
   - (B) a line
   - (C) a parabola
   - (D) a midsegment

9. *Multiple Choice* What is a focus?
   - (A) a point
   - (B) a line
   - (C) a parabola
   - (D) a midsegment

Algebra 1  Standardized Test Practice 12.5  **85**

NAME _____ CLASS _____ DATE _____

# Standardized Test Practice
## 12.6 The Tangent Function

**TEST TAKING STRATEGY** Be aware of similar answers.

1. **Multiple Choice** In a right triangle, what is the ratio used to represent the tangent of an angle?

   Ⓐ $\dfrac{\text{opposite}}{\text{hypotenuse}}$  Ⓑ $\dfrac{\text{adjacent}}{\text{hypotenuse}}$

   Ⓒ $\dfrac{\text{opposite}}{\text{adjacent}}$  Ⓓ $\dfrac{\text{adjacent}}{\text{opposite}}$

2. **Multiple Choice** The tangent ratio is a fundamental part of what branch of mathematics?

   Ⓐ geometry
   Ⓑ trigonometry
   Ⓒ algebra
   Ⓓ Calculus

3. **Multiple Choice** The tangent ratio is a function of which of the following?

   Ⓐ triangle size
   Ⓑ the domain
   Ⓒ the range
   Ⓓ angle measure

4. **Multiple Choice** Which ratio defines tan $A$ in the following diagram?

   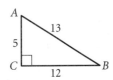

   Ⓐ $\dfrac{5}{12}$  Ⓑ $\dfrac{5}{13}$

   Ⓒ $\dfrac{12}{13}$  Ⓓ $\dfrac{12}{5}$

*Quantitative Comparison* In Exercises 5–7, choose the letter of the statement below that is true about the quantities in Columns I and II.

**A** The number in Column I is greater.
**B** The number in Column II is greater.
**C** The two numbers are equal.
**D** The relationship cannot be determined from the given information.

| | Column I | Column II |
|---|---|---|
| 5. | tan 56° | tan 48° |
| | Ⓐ Ⓑ | Ⓒ Ⓓ |
| 6. | the measure of angle $A$ if $\tan A = \dfrac{5}{12}$ | the measure of angle $A$ if $\tan A = \dfrac{30{,}000}{184{,}800}$ |
| | Ⓐ Ⓑ | Ⓒ Ⓓ |

7.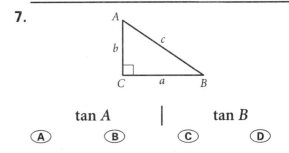

   | tan $A$ | tan $B$ |
   |---|---|
   | Ⓐ Ⓑ | Ⓒ Ⓓ |

8. **Multiple Choice** At a point 200 meters from the base of a building, the angle of elevation is 62°. What is the height of the building to the nearest meter?

   Ⓐ 426 meters
   Ⓑ 177 meters
   Ⓒ 376 meters
   Ⓓ 156 meters

NAME _____ CLASS _____ DATE _____

# Standardized Test Practice
## 12.7 The Sine and Cosine Functions

**TEST TAKING STRATEGY** Treat multiple choice questions like they are true/false questions.

**1. Multiple Choice** Which of the following represents the sine function?

Ⓐ $\dfrac{\text{length of the leg opposite } \angle A}{\text{length of the leg adjacent to } \angle A}$

Ⓑ $\dfrac{\text{length of the leg opposite } \angle A}{\text{length of the hypotenuse}}$

Ⓒ $\dfrac{\text{length of the leg adjacent to } \angle A}{\text{length of the hypotenuse}}$

Ⓓ none of the above

**2. Multiple Choice** What is the approximate length of side $x$ in $\triangle XYZ$?

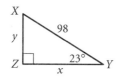

Ⓐ 41.6  Ⓑ 90.2
Ⓒ 106.5  Ⓓ 38.3

**3. Multiple Choice** What is the angle measure, to the nearest degree, that has a sine value of 0.129?

Ⓐ 7°  Ⓑ 1°
Ⓒ 83°  Ⓓ 0°

**4. Multiple Choice** What function would be used to find the value of $x$?

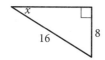

Ⓐ sine
Ⓑ cosine
Ⓒ tangent
Ⓓ Pythagorean Theorem

***Quantitative Comparison*** In Exercises 5–7, choose the letter of the statement below that is true about the quantities in Columns I and II.

**A** The number in Column I is greater.
**B** The number in Column II is greater.
**C** The two numbers are equal.
**D** The relationship cannot be determined from the given information.

|   | Column I | Column II |
|---|---|---|
| 5. | sin 56° | cos 56° |
|   | Ⓐ   Ⓑ   Ⓒ   Ⓓ | |
| 6. | the measure of angle $A$ if $\cos A = \dfrac{3}{4}$ | the measure of angle $A$ if $\sin A = \dfrac{59}{223}$ |
|   | Ⓐ   Ⓑ   Ⓒ   Ⓓ | |

7.
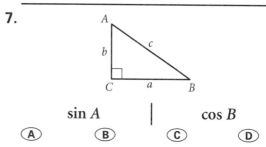

sin $A$ | cos $B$

Ⓐ   Ⓑ   Ⓒ   Ⓓ

**8. Multiple Choice** A wire is fastened to the top of a TV tower 40 feet above the ground, and forms an angle of 52° with the tower. How long is the wire to the nearest foot?

Ⓐ 25 feet
Ⓑ 51 feet
Ⓒ 65 feet
Ⓓ 32 feet

Algebra 1

NAME _____ CLASS _____ DATE _____

# Standardized Test Practice
## 12.8 Introduction to Matrices

**TEST TAKING STRATEGY** Be sure you understand what is being asked in the question.

1. **Multiple Choice** Which grouping symbol is used with matrices?
   - Ⓐ braces
   - Ⓑ parentheses
   - Ⓒ brackets
   - Ⓓ quotations

2. **Multiple Choice** When are two matrices equal?
   - Ⓐ When their dimensions are the same.
   - Ⓑ When their corresponding entries are equal.
   - Ⓒ both A and B
   - Ⓓ neither A nor B

3. **Multiple Choice** What is the sum of
   $\begin{bmatrix} 2 & -3 & 5 \\ 7 & 0 & -\frac{1}{2} \end{bmatrix} + \begin{bmatrix} 6 & 7 & -8 \\ -1 & \frac{2}{3} & \frac{5}{2} \end{bmatrix}$?

   - Ⓐ $\begin{bmatrix} 8 & -10 & -3 \\ 8 & 0 & 2 \end{bmatrix}$
   - Ⓑ $\begin{bmatrix} 8 & 4 & -3 \\ 6 & \frac{2}{3} & 2 \end{bmatrix}$
   - Ⓒ $\begin{bmatrix} 8 & -4 & 3 \\ 6 & 0 & 3 \end{bmatrix}$
   - Ⓓ $\begin{bmatrix} 8 & -3 & -3 \\ 8 & \frac{2}{3} & 2 \end{bmatrix}$

4. **Multiple Choice** What is the identity matrix for $\begin{bmatrix} 3 & 4 & 5 \\ 6 & 7 & 8 \\ 4 & 3 & 3 \end{bmatrix}$?

   - Ⓐ $\begin{bmatrix} 1 & 0 & 0 \\ 1 & 0 & 0 \\ 1 & 0 & 0 \end{bmatrix}$
   - Ⓑ $\begin{bmatrix} 1 & 1 & 1 \\ 0 & 0 & 0 \\ 0 & 0 & 0 \end{bmatrix}$
   - Ⓒ $\begin{bmatrix} 1 & 0 & 0 \\ 0 & 1 & 0 \\ 0 & 0 & 1 \end{bmatrix}$
   - Ⓓ $\begin{bmatrix} 0 & 0 & 1 \\ 0 & 1 & 0 \\ 1 & 0 & 0 \end{bmatrix}$

**Quantitative Comparison** In Exercises 5–7, choose the letter of the statement below that is true about the quantities in Columns I and II.

A The number in Column I is greater.
B The number in Column II is greater.
C The two numbers are equal.
D The relationship cannot be determined from the given information.

Use matrices $M$ and $N$ for Questions 5–7. Assume that $M = N$.

$M = \begin{bmatrix} 3(a-2) & 56 & -\frac{2}{3}c \\ (-3d-4) & 42 & \frac{1}{2}f \end{bmatrix}$

$N = \begin{bmatrix} -15 & -7b & -9 \\ -(2-d) & 0.4e & \left(\frac{2}{3}f - 5\right) \end{bmatrix}$

| | Column I | Column II |
|---|---|---|
| 5. | $a$ | $b$ |
|   | Ⓐ Ⓑ | Ⓒ Ⓓ |
| 6. | $c$ | $d$ |
|   | Ⓐ Ⓑ | Ⓒ Ⓓ |
| 7. | $e$ | $f$ |
|   | Ⓐ Ⓑ | Ⓒ Ⓓ |

8. **Multiple Choice** The first row of the first column of the resulting matrix is:

   $\begin{bmatrix} 2 \\ 5 \\ -6 \end{bmatrix} - \begin{bmatrix} 1 & 2 & 3 \end{bmatrix}$

   - Ⓐ positive
   - Ⓑ 0
   - Ⓒ negative
   - Ⓓ not possible

# SAT/ACT Chapter Test
## Chapter 12  Radicals, Functions, and Coordinate Geometry

**TEST TAKING STRATEGY**  Begin with the questions you know you can answer correctly.

1. **Multiple Choice**  What is the simplest radical form for the radical expression $\sqrt{6}(3 - \sqrt{3})$?
   - Ⓐ 0
   - Ⓑ $3\sqrt{6} - \sqrt{18}$
   - Ⓒ $6\sqrt{3} - \sqrt{18}$
   - Ⓓ $3\sqrt{6} - 3\sqrt{2}$

2. **Multiple Choice**  What is the distance between the points $(-5, 6)$ and $(5, -3)$?
   - Ⓐ $d = 3$
   - Ⓑ $d = \sqrt{181}$
   - Ⓒ $d = \sqrt{109}$
   - Ⓓ $d = 9$

3. **Multiple Choice**  What is the exact solution to the equation $\sqrt{x^2 - 4x + 4} = x - 4$?
   - Ⓐ $x = 3$
   - Ⓑ $x = -2$
   - Ⓒ $x = -3$
   - Ⓓ $x = 4$

4. **Multiple Choice**  What is the length of the missing side of a right triangle if the measure of one leg is 6 centimeters and the hypotenuse measures 18 centimeters?
   - Ⓐ $6\sqrt{10}$ centimeters
   - Ⓑ 12 centimeters
   - Ⓒ $12\sqrt{2}$ centimeters
   - Ⓓ 6 centimeters

5. **Multiple Choice**  Find the sum.
   $2\begin{bmatrix} -3 & 1 \\ 0 & -4 \end{bmatrix} + \begin{bmatrix} 4 & -2 \\ -5 & 7 \end{bmatrix}$
   - Ⓐ $\begin{bmatrix} 1 & -1 \\ -5 & 3 \end{bmatrix}$
   - Ⓑ $\begin{bmatrix} -2 & 0 \\ -5 & 3 \end{bmatrix}$
   - Ⓒ $\begin{bmatrix} -2 & -1 \\ -5 & 3 \end{bmatrix}$
   - Ⓓ $\begin{bmatrix} -2 & 0 \\ -5 & -1 \end{bmatrix}$

**Quantitative Comparison**  In Exercises 6–8, choose the letter of the statement below that is true about the quantities in Columns I and II.

- **A** The number in Column I is greater.
- **B** The number in Column II is greater.
- **C** The two numbers are equal.
- **D** The relationship cannot be determined from the given information.

| | Column I | Column II |
|---|---|---|
| 6. | $\tan 38°$ | $\sin 24°$ |
| | Ⓐ  Ⓑ | Ⓒ  Ⓓ |
| 7. | the radius of the circle represented by $x^2 + (y + 3)^2 = 27$ | the radius of the circle represented by $(x - 4)^2 + y^2 = 10.3$ |
| | Ⓐ  Ⓑ | Ⓒ  Ⓓ |
| 8. | $\sqrt{6.25}$ | $\sqrt{\dfrac{25}{4}}$ |
| | Ⓐ  Ⓑ | Ⓒ  Ⓓ |

9. **Multiple Choice**  What is the equation of the circle whose center is $(5, -3)$ and whose radius is $2\sqrt{3}$?
   - Ⓐ $(x - 5)^2 + (x + 3)^2 = 12$
   - Ⓑ $(x - 5)^2 + (x - 3)^2 = 36$
   - Ⓒ $(x - 5)^2 + (x + 3)^2 = 36$
   - Ⓓ $(x + 5)^2 + (x - 3)^2 = 12$

10. **Multiple Choice**  Which of the following lengths of the sides of a triangle can *not* form a right triangle?
    - Ⓐ $4, 3\sqrt{2}, \sqrt{34}$
    - Ⓑ $\sqrt{3}, \sqrt{5}, 2\sqrt{2}$
    - Ⓒ $3, 4, 5$
    - Ⓓ $1, 1, 2$

# Standardized Test Practice
## 13.1 Theoretical Possibility

**TEST TAKING STRATEGY** Making a diagram or a sketch may help you understand the question.

1. **Multiple Choice** A coin is tossed and a number cube is rolled. What is the sample space for the experiment?

   Ⓐ {(H, 1), (H, 2), (H, 3), (H, 4), (H, 5), (H, 6)}
   Ⓑ {H, T, 1, 2, 3, 4, 5, 6}
   Ⓒ {(H, 1), (H, 2), (H, 3), (H, 4), (H, 5), (H, 6), (T, 1), (T, 2), (T, 3), (T, 4), (T, 5), (T, 6)}
   Ⓓ {(T, 1), (T, 2), (T, 3), (T, 4), (T, 5), (T, 6)}

2. **Multiple Choice** A number cube is rolled twice. What are the favorable outcomes for a sum of 8?

   Ⓐ (2, 6), (3, 5)
   Ⓑ (2, 6), (3, 5), (4, 4)
   Ⓒ (2, 6), (6, 2), (3, 5), (5, 3)
   Ⓓ (2, 6), (6, 2), (3, 5), (5, 3), (4, 4)

3. **Multiple Choice** A circular spinner is divided into 4 equal-sized sectors. Two are blue, one is red, and one is white. What is the probability of landing on white?

   Ⓐ $\frac{1}{4}$  Ⓑ $\frac{1}{2}$
   Ⓒ $\frac{1}{3}$  Ⓓ 0

4. **Multiple Choice** If you select a letter at random from the English alphabet, what is the probability that it will be in the word *probability*?

   Ⓐ $\frac{11}{26}$  Ⓑ $\frac{9}{26}$
   Ⓒ $\frac{1}{3}$  Ⓓ $\frac{4}{13}$

**Quantitative Comparison** In Exercises 5–7, choose the letter of the statement below that is true about the quantities in Columns I and II.

**A** The number in Column I is greater.
**B** The number in Column II is greater.
**C** The two numbers are equal.
**D** The relationship cannot be determined from the given information.

| Column I | Column II |
|---|---|
| 5. the probability of a sum of 5 when a number cube is rolled twice | the probability of a sum of 6 when a number cube is rolled twice |
| Ⓐ  Ⓑ | Ⓒ  Ⓓ |
| 6. the probability of a sum of 2 when a number cube is rolled twice | the probability of a sum of 12 when a number cube is rolled twice |
| Ⓐ  Ⓑ | Ⓒ  Ⓓ |
| 7. the probability of selecting an even integer if one is selected at random from the integers 1 through 10 | the probability of tossing one head and one tail if a coin is tossed twice |
| Ⓐ  Ⓑ | Ⓒ  Ⓓ |

8. **Multiple Choice** Consider all the possible arrangements of the letters in the word *now*. What is the probability that an arrangement chosen at random is a word?

   Ⓐ $\frac{1}{6}$  Ⓑ $\frac{1}{3}$
   Ⓒ $\frac{1}{4}$  Ⓓ $\frac{1}{2}$

# Standardized Test Practice
## 13.2 Counting the Elements of a Set

**TEST TAKING STRATEGY** Make an inference to fill in missing information.

1. **Multiple Choice** Which best describes the intersection of two disjoint sets?
   - Ⓐ the elements that are in both sets
   - Ⓑ all elements of both sets
   - Ⓒ overlapping circles
   - Ⓓ non-overlapping circles

2. **Multiple Choice** The face cards in a 52-card deck are the jacks, queens, and kings. How many cards are red face cards?
   - Ⓐ 26
   - Ⓑ 6
   - Ⓒ 4
   - Ⓓ 16

Use the table below for Exercises 3–5.

Survey of the Number of College Students Studying a Foreign Language

|  | Yes | No |
|---|---|---|
| First year | 212 | 600 |
| Second year | 400 | 395 |
| Third year | 235 | 450 |
| Fourth year | 180 | 495 |

3. **Multiple Choice** How many first-year OR second-year students are studying a foreign language?
   - Ⓐ 188
   - Ⓑ 1617
   - Ⓒ 612
   - Ⓓ 388

4. **Multiple Choice** How many third-year OR fourth-year students were surveyed?
   - Ⓐ 1360
   - Ⓑ 945
   - Ⓒ 415
   - Ⓓ 530

5. **Multiple Choice** How many students are studying a foreign language OR are third-year students?
   - Ⓐ 1712
   - Ⓑ 1477
   - Ⓒ 1027
   - Ⓓ 685

**Quantitative Comparison** In Exercises 6–8, choose the letter of the statement below that is true about the quantities in Columns I and II.

- **A** The number in Column I is greater.
- **B** The number in Column II is greater.
- **C** The two numbers are equal.
- **D** The relationship cannot be determined from the given information.

| | Column I | Column II |
|---|---|---|
| 6. | the probability that an integer from 1 through 20 is a multiple of 3 AND a multiple of 5 | the probability that an integer from 1 through 20 is a multiple of 3 OR a multiple of 5 |
| | Ⓐ   Ⓑ | Ⓒ   Ⓓ |
| 7. | the probability of drawing a face card at random from a 52-card deck | the probability of drawing an ace OR a ten at random from a 52-card deck |
| | Ⓐ   Ⓑ | Ⓒ   Ⓓ |
| 8. | the number of even integers in the set of numbers from 1 through 100 | the number of odd integers in the set of numbers from 1 through 100 |
| | Ⓐ   Ⓑ | Ⓒ   Ⓓ |

9. **Multiple Choice** What is the probability of drawing a club OR a king at random from a 52-card deck?
   - Ⓐ $\frac{4}{13}$
   - Ⓑ $\frac{15}{52}$
   - Ⓒ $\frac{17}{52}$
   - Ⓓ $\frac{1}{4}$

NAME _____ CLASS _____ DATE _____

# Standardized Test Practice
## 13.3 The Fundamental Counting Principle

**TEST TAKING STRATEGY** Underline the key words and numbers in the question.

The tree diagram shows the possible outcomes when a coin is tossed 3 times. Use the diagram for Exercises 1–2.

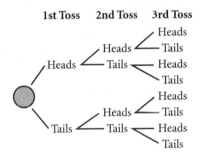

1. **Multiple Choice** How many outcomes have exactly 2 heads?
   - Ⓐ 6
   - Ⓑ 3
   - Ⓒ 7
   - Ⓓ 2

2. **Multiple Choice** How many outcomes are there in all?
   - Ⓐ 2
   - Ⓑ 4
   - Ⓒ 6
   - Ⓓ 8

3. **Multiple Choice** A dairy produces 1% milk, 2% milk, skim milk, and whole milk. The milk is sold in quart, half gallon, and gallon containers. Which number reflects the number of milk choices?
   - Ⓐ 4
   - Ⓑ 12
   - Ⓒ 3
   - Ⓓ 7

4. **Multiple Choice** A store has jerseys for 12 basketball teams. The jerseys come in small, medium, large, and extra large. How many different jerseys are there?
   - Ⓐ 12
   - Ⓑ 16
   - Ⓒ 48
   - Ⓓ 36

**Quantitative Comparison** In Exercises 5–7, choose the letter of the statement below that is true about the quantities in Columns I and II.

- **A** The number in Column I is greater.
- **B** The number in Column II is greater.
- **C** The two numbers are equal.
- **D** The relationship cannot be determined from the given information.

| | Column I | Column II |
|---|---|---|
| 5. | the number of ways of drawing a heart from one 52-card deck AND a face card from another deck | the number of ways of choosing 1 CD AND 1 tape from a collection of 10 CDs and 10 tapes |
| | Ⓐ  Ⓑ | Ⓒ  Ⓓ |
| 6. | the number of 5-digit numbers that can be formed from the digits 0 through 9 if the digits cannot repeat | the number of 5-digit numbers that can be formed from the digits 0 through 9 if the digits can repeat |
| | Ⓐ  Ⓑ | Ⓒ  Ⓓ |
| 7. | the number of sweater-skirt outfits from 6 sweaters and 4 skirts | the number of three-piece outfits from 2 jackets, 4 shirts, and 3 pairs of slacks |
| | Ⓐ  Ⓑ | Ⓒ  Ⓓ |

8. **Multiple Choice** In how many ways can 5 people arrange themselves in seats 22 through 26 of row J?
   - Ⓐ 120
   - Ⓑ 625
   - Ⓒ 125
   - Ⓓ 3125

92   Standardized Test Practice 13.3   Algebra 1

# Standardized Test Practice
## 13.4 Independent Events

**TEST TAKING STRATEGY** Restate the question to determine if you answered it correctly.

1. **Multiple Choice** A red and a green number cube are rolled. What is the probability that the red cube shows a number greater than 3 AND the green cube shows a number less than or equal to 3?
   - (A) $\frac{1}{6}$
   - (B) $\frac{1}{4}$
   - (C) $\frac{1}{3}$
   - (D) $\frac{1}{2}$

2. **Multiple Choice** A circular spinner has equal-sized sectors labeled 1 through 8. If the spinner is spun twice, what is the probability of spinning a 6 AND a number less than 5?
   - (A) $\frac{1}{2}$
   - (B) $\frac{1}{8}$
   - (C) $\frac{1}{4}$
   - (D) $\frac{1}{16}$

3. **Multiple Choice** If a coin is tossed 3 times, what is the probability of it landing heads up 3 times?
   - (A) $\frac{1}{3}$
   - (B) $\frac{1}{2}$
   - (C) $\frac{1}{8}$
   - (D) $\frac{1}{4}$

4. **Multiple Choice** A marble is drawn from a bag containing 6 white, 4 red, and 4 blue marbles. One marble is drawn, then replaced, and a second one is drawn. What is the probability of drawing a red marble AND a white marble?
   - (A) $\frac{3}{7}$
   - (B) $\frac{6}{49}$
   - (C) $\frac{2}{7}$
   - (D) $\frac{5}{7}$

**Quantitative Comparison** In Exercises 5–6, choose the letter of the statement below that is true about the quantities in Columns I and II.

- **A** The number in Column I is greater.
- **B** The number in Column II is greater.
- **C** The two numbers are equal.
- **D** The relationship cannot be determined from the given information.

| | Column I | Column II |
|---|---|---|
| 5. | the probability of getting a head first AND a tail second on 2 tosses of a coin | the probability of getting an even number when a number cube is rolled |
| | (A) (B) | (C) (D) |
| 6. | the probability of getting a sum greater than or equal to 10 when two number cubes are rolled | the probability of getting a sum less than or equal to 4 when two number cubes are rolled |
| | (A) (B) | (C) (D) |

7. **Multiple Choice** Two airplanes are on time 90% of the time. What is the probability that both will be on time the next time they fly?
   - (A) 90%
   - (B) 180%
   - (C) 50%
   - (D) 81%

NAME _____ CLASS _____ DATE _____

# Standardized Test Practice
## 13.5 Simulations

**TEST TAKING STRATEGY** Do first those exercises you know you can answer correctly.

1. **Multiple Choice** There is about a 33% chance of rain on Sunday. Which could be used to represent rain on Sunday?
   - Ⓐ heads when a coin is tossed
   - Ⓑ the slips numbered 1 and 2 from slips of paper numbered 1 through 10
   - Ⓒ drawing a heart from a deck of cards
   - Ⓓ the faces numbered 1 and 2 on a number cube

Random drawings of slips of paper numbered 1 through 10 were used to generate the numbers shown in the table below. Use the table for Exercises 2–4.

| Trial  | 1  | 2  | 3  | 4  | 5  |
|--------|----|----|----|----|----|
| Number | 10 | 10 | 2  | 6  | 5  |
| Trial  | 6  | 7  | 8  | 9  | 10 |
| Number | 8  | 1  | 4  | 10 | 3  |
| Trial  | 11 | 12 | 13 | 14 | 15 |
| Number | 8  | 10 | 3  | 4  | 1  |
| Trial  | 16 | 17 | 18 | 19 | 20 |
| Number | 10 | 2  | 1  | 6  | 9  |

2. **Multiple Choice** If the numbers 1 through 5 represent True and the numbers 6 through 10 represent False, what percent of the choices would you expect to be true in a simulation of a True-False test with 20 questions?
   - Ⓐ 10%    Ⓑ 20%
   - Ⓒ 50%    Ⓓ 55%

3. **Multiple Choice** Suppose each number represents a percentage score on a test: 1 represents 10%, 2 represents 20%, and so on. If 40 students took the test, how many would you expect to have a score of 80% or higher?
   - Ⓐ 16    Ⓑ 8
   - Ⓒ 10    Ⓓ 12

**Quantitative Comparison** In Exercises 4–6, choose the letter of the statement below that is true about the quantities in Columns I and II.

**A** The number in Column I is greater.
**B** The number in Column II is greater.
**C** The two numbers are equal.
**D** The relationship cannot be determined from the given information.

| | Column I | Column II |
|---|---|---|
| 4. | the number of random digits needed to simulate the selection of 1 of 5 possible responses to a multiple-choice item | the number of random digits needed to simulate the selection of a day of the week |
| | Ⓐ   Ⓑ | Ⓒ   Ⓓ |
| 5. | the number of expected favorable outcomes in 100 trials if the experimental probability is 25% | the number of expected favorable outcomes in 200 trials if the experimental probability is 12.5% |
| | Ⓐ   Ⓑ | Ⓒ   Ⓓ |
| 6. | the number of trials need to result in a probability of 75% | the number of trials needed to result in a probability of 50% |
| | Ⓐ   Ⓑ | Ⓒ   Ⓓ |

7. **Multiple Choice** The experimental probability of an event is 0.5%. If the experiment consisted of 200 trials, how many favorable outcomes can you expect?
   - Ⓐ 1    Ⓑ 10
   - Ⓒ 5    Ⓓ 2

94  Standardized Test Practice 13.5  Algebra 1

# SAT/ACT Chapter Test
## Chapter 13   Probability

**TEST TAKING STRATEGY**   Always look at all of the answer choices before choosing one.

1. **Multiple Choice**  A spinner has 3 equal-sized sectors labeled 1, 2, and 3. Ty spins the spinner 3 times. Which outcomes in the sample space have 1 as the first number?
   - Ⓐ {(1, 2, 3)}
   - Ⓑ {(1, 1, 1), (1, 1, 2), (1, 1, 3), (1, 2, 1), (1, 2, 2), (1, 2, 3), (1, 3, 1), (1, 3, 2), (1, 3, 3)}
   - Ⓒ {(1, 1, 1), (1, 2, 2), (1, 2, 3), (1, 3, 3)}
   - Ⓓ {(1, 1, 1)}

2. **Multiple Choice**  What is the probability that a card chosen at random from a deck of 52 cards will be a 10?
   - Ⓐ $\frac{5}{26}$
   - Ⓑ $\frac{1}{52}$
   - Ⓒ $\frac{1}{10}$
   - Ⓓ $\frac{1}{13}$

3. **Multiple Choice**  Which integers from 1 through 15 are odd AND multiples of 5?
   - Ⓐ 5, 15
   - Ⓑ 1, 3, 5, 7, 9, 11, 13, 15
   - Ⓒ 1, 3, 5, 7, 9, 10, 11, 13, 15
   - Ⓓ 5, 10, 15

4. **Multiple Choice**  A yellow number cube and a blue number cube are rolled together. What is the probability that the yellow cube is odd AND the blue cube is less than 5?
   - Ⓐ $\frac{5}{36}$
   - Ⓑ $\frac{5}{18}$
   - Ⓒ $\frac{11}{36}$
   - Ⓓ $\frac{1}{3}$

5. **Multiple Choice**  How many ways can a triangle, with vertices P, Q, and R, be named?
   - Ⓐ 27
   - Ⓑ 6
   - Ⓒ 9
   - Ⓓ 1

**Quantitative Comparison**  In Exercises 6–8, choose the letter of the statement below that is true about the quantities in Columns I and II.

- **A**  The number in Column I is greater.
- **B**  The number in Column II is greater.
- **C**  The two numbers are equal.
- **D**  The relationship cannot be determined from the given information.

| | Column I | Column II |
|---|---|---|
| 6. | the number of outcomes when a coin is flipped and a number cube is rolled | the number of ways of choosing 1 of 4 T-shirts and 1 of 3 pairs of jeans |
| | Ⓐ   Ⓑ | Ⓒ   Ⓓ |
| 7. | the probability of drawing 2 aces if 2 cards are drawn with replacement from a 52-card deck | the probability of drawing 2 spades if 2 cards are drawn with replacement from a 52-card deck |
| | Ⓐ   Ⓑ | Ⓒ   Ⓓ |
| 8. | the probability that both numbers are less than 4 if one is drawn from {2, 4, 6, 8} and the other from {1, 2, 3, 4} | the probability that both numbers are 4 if one is drawn from {2, 4, 6, 8} and the other from {1, 2, 3, 4} |
| | Ⓐ   Ⓑ | Ⓒ   Ⓓ |

9. **Multiple Choice**  The experimental probability of an event is 12.5%. If the experiment consisted of 200 trials, how many favorable outcomes are there?
   - Ⓐ 20
   - Ⓑ 125
   - Ⓒ 25
   - Ⓓ 8

NAME _____ CLASS _____ DATE _____

# Standardized Test Practice
## 14.1 Graphing Functions and Relations

**TEST TAKING STRATEGY** Read directions carefully.

1. **Multiple Choice** What function is represented by the diagram shown?

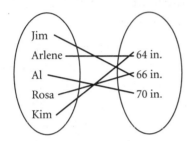

   Ⓐ {(Jim, Rosa, 66), (Arlene, Kim, 64), (Al, 70)}
   Ⓑ {(Jim, 66), (Arlene, 64), (Al, 70)}
   Ⓒ {(Jim, 66), (Arlene, 64), (Al, 70), (Rosa, 66), (Kim, 64)}
   Ⓓ {(Jim, 64), (Arlene, 64), (Al, 65), (Rosa, 70), (Kim, 70)}

2. **Multiple Choice** If $f(x) = 3x^2 - 5$, what is $f(-1)$?

   Ⓐ $-2$    Ⓑ $-8$
   Ⓒ $3$     Ⓓ $-5$

3. **Multiple Choice** If $f(3) = 0$, what is $f(x)$?

   Ⓐ $f(x) = x^2 - x + 3$
   Ⓑ $f(x) = 2x^2 - 4x - 6$
   Ⓒ $f(x) = 3x - 3$
   Ⓓ $f(x) = -3$

4. **Multiple Choice** If the domain of $g(x) = -2|x| + 4$ is the set of integers $-1, 0,$ and $1$, which is a listing of the ordered pairs in this function?

   Ⓐ {(−1, 6), (0, 4), (1, 2)}
   Ⓑ {(−1, 2), (0, 0), (1, −2)}
   Ⓒ {(−1, 2), (0, 4), (1, 2)}
   Ⓓ {(−1, −2), (0, 4), (1, −2)}

**Quantitative Comparison** In Exercises 5–7, choose the letter of the statement below that is true about the quantities in Columns I and II.

A The number in Column I is greater.
B The number in Column II is greater.
C The two numbers are equal.
D The relationship cannot be determined from the given information.

| | Column I | | Column II | |
|---|---|---|---|---|
| 5. | $f(-1)$ if $f(x) = \|3x - 4\|$ | | $f(0)$ if $f(x) = 3x^2 + 7$ | |
| | Ⓐ | Ⓑ | Ⓒ | Ⓓ |
| 6. | $g(2)$ if $g(x) = x^2 - x$ | | $g(4)$ if $g(x) = -\|x\|$ | |
| | Ⓐ | Ⓑ | Ⓒ | Ⓓ |
| 7. | $h(-5)$ if $h(x) = 3x + 3$ | | $h(1)$ if $h(x) = 2x^2 - 10$ | |
| | Ⓐ | Ⓑ | Ⓒ | Ⓓ |

8. **Multiple Choice** Which of the following is the parent function for the family of rational functions?

   Ⓐ $f(x) = \dfrac{3}{x^2}$

   Ⓑ $f(x) = \dfrac{1}{2}x^2$

   Ⓒ $f(x) = \sqrt{x}$

   Ⓓ $f(x) = \dfrac{1}{x}$

NAME _____ CLASS _____ DATE _____

# Standardized Test Practice
## 14.2 Translations

**TEST TAKING STRATEGY** Restate each question to verify that you understand it.

1. **Multiple Choice** A horizontal translation of 5 units to the right of the graph of $y = x^2$ can be written as:

   Ⓐ $y = (x - 5)^2$
   Ⓑ $y = x^2 - 5$
   Ⓒ $y = (x + 5)^2$
   Ⓓ $y = x^2 + 5$

2. **Multiple Choice** Which graph is of the function $y = |x + 1| - 1$?

   Ⓐ 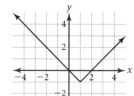   Ⓑ

   Ⓒ 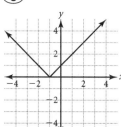   Ⓓ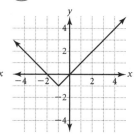

3. **Multiple Choice** The graph is a translation of $y = x^2$. Describe the function.

   Ⓐ $y = (x + 3)^2 - 2$
   Ⓑ $y = (x - 3)^2 - 2$
   Ⓒ $y = (x - 3)^2 + 2$
   Ⓓ $y = (x + 3)^2 + 2$

*Quantitative Comparison* In Exercises 4–6, choose the letter of the statement below that is true about the quantities in Columns I and II.

**A** The number in Column I is greater.
**B** The number in Column II is greater.
**C** The two numbers are equal.
**D** The relationship cannot be determined from the given information.

| | Column I | Column II |
|---|---|---|
| 4. | the value of $h$ in $y = f(x - h)$ if the graph of $y = f(x)$ is translated to the right | the value of $k$ in $y = f(x) + k$ if the graph of $y = f(x)$ is translated up |
| | Ⓐ    Ⓑ | Ⓒ    Ⓓ |
| 5. | the value of $h$ in $y = f(x - h)$ if the graph of $y = f(x)$ is translated to the right 3 units | the value of $k$ in $y = f(x) + k$ if the graph of $y = f(x)$ is translated up 3 units |
| | Ⓐ    Ⓑ | Ⓒ    Ⓓ |
| 6. | the value of the $y$-coordinate when (0, 0) is translated | |
| | left 4 units | right 4 units |
| | Ⓐ    Ⓑ | Ⓒ    Ⓓ |

7. **Multiple Choice** The graph of $f(x) = x^2$ is translated left 3 units and down 2 units. Which equation describes the translated graph?

   Ⓐ $f(x) = (x - 3)^2 - 2$
   Ⓑ $f(x) = x^2 - 5$
   Ⓒ $f(x) = (x + 3)^2 - 2$
   Ⓓ $f(x) = (x + 3)^2 + 2$

Algebra 1                    Standardized Test Practice 14.2    97

NAME _____ CLASS _____ DATE _____

# Standardized Test Practice
## 14.3 Stretches and Compressions

**TEST TAKING STRATEGY** Use context to help define an unknown word.

1. **Multiple Choice** The graph of $y = x^2$ is compressed horizontally by a factor of $\frac{1}{5}$. Which function represents this transformation?

   Ⓐ $y = \frac{1}{5}x^2$     Ⓑ $y = 5x^2$

   Ⓒ $y = \left(\frac{1}{5}x\right)^2$     Ⓓ $y = (5x)^2$

2. **Multiple Choice** The graph of $y = \sqrt{x}$ is stretched vertically by a factor of 1.5. Which function describes this transformation?

   Ⓐ $y = \sqrt{x} + 1.5$

   Ⓑ $y = \sqrt{1.5x}$

   Ⓒ $y = 1.5\sqrt{x}$

   Ⓓ $y = \frac{\sqrt{x}}{1.5}$

3. **Multiple Choice** Which function describes a horizontal compression that results in the same graph as the vertical stretch described in $y = 1.44x^2$?

   Ⓐ $y = (1.2x)^2$    Ⓑ $y = \left(\frac{x}{1.2}\right)^2$

   Ⓒ $y = 1.2x^2$    Ⓓ $y = \frac{x^2}{12}$

4. **Multiple Choice** If the graph of $y = f(bx)$ is a horizontal stretch of the graph of $y = f(x)$, then what must be true of $b$?

   Ⓐ $b < 0$     Ⓑ $0 < b < 1$
   Ⓒ $b > 1$     Ⓓ $b \geq 0$

**Quantitative Comparison** In Exercises 5–7, choose the letter of the statement below that is true about the quantities in Columns I and II.

A The number in Column I is greater.
B The number in Column II is greater.
C The two numbers are equal.
D The relationship cannot be determined from the given information.

| | Column I | Column II |
|---|---|---|
| 5. | the y-coordinate of the point $(1, y)$ on the graph of $y = 3x^2$ | $y = |3x|$ |
| | Ⓐ   Ⓑ   Ⓒ   Ⓓ | |
| 6. | the value of $a$ if the graph of $y = af(x)$ is a vertical stretch of the graph of $y = f(x)$ | the value of $b$ if the graph of $y = f(bx)$ is a horizontal compression of the graph of $y = f(x)$ |
| | Ⓐ   Ⓑ   Ⓒ   Ⓓ | |
| 7. | the value of $b$ if the graph of $y = f(bx)$ is a horizontal compression of the graph of $y = f(x)$ | the value of $a$ if the graph of $y = af(x)$ is a vertical compression of the graph of $y = f(x)$ |
| | Ⓐ   Ⓑ   Ⓒ   Ⓓ | |

8. **Multiple Choice** Which is a vertical compression of the graph of $y = x^2$?

   Ⓐ $y = \frac{x^2}{10}$     Ⓑ $y = 3x^2$

   Ⓒ $y = -x^2$     Ⓓ $y = \frac{x^2}{0.2}$

NAME _____ CLASS _____ DATE _____

# Standardized Test Practice
## 14.4 Reflections

**TEST TAKING STRATEGY** Look closely at each choice, be aware of similar answers.

1. **Multiple Choice** Which equation describes the graph of $f(x) = x + 2$ reflected across the x-axis?

   Ⓐ $f(x) = -x - 2$  Ⓑ $f(x) = -x + 2$
   Ⓒ $f(x) = x - 2$   Ⓓ $f(x) = -x$

2. **Multiple Choice** How does the graph of $y = |-x|$ compare to the graph of $y = |x|$?

   Ⓐ It is a reflection across the x-axis.
   Ⓑ It is a translation 1 unit to the right.
   Ⓒ It is identical.
   Ⓓ It is a horizontal stretch by a factor of 1.

3. **Multiple Choice** The point $(-3, -2)$ is on the graph of a function before it is reflected across the y-axis. Which of the following coordinate pairs represent one point on the reflection?

   Ⓐ $(-3, 2)$   Ⓑ $(3, -2)$
   Ⓒ $(3, 2)$    Ⓓ $(0, 0)$

4. **Multiple Choice** Which is an equation for the reflection across the y-axis of the graph of $y = x^2 + 3x$?

   Ⓐ $y = x^2 - 3x$    Ⓑ $y = -x^2 - 3x$
   Ⓒ $y = -x^2 + 3x$   Ⓓ $y = (-x)^2$

5. **Multiple Choice** The graph of $y = 2x - 4$ is reflected across the x-axis. Which point is on both graphs?

   Ⓐ $(0, 4)$   Ⓑ $(0, -4)$
   Ⓒ $(2, 0)$   Ⓓ $(0, 0)$

*Quantitative Comparison* In Exercises 6–8, choose the letter of the statement below that is true about the quantities in Columns I and II.

**A** The number in Column I is greater.
**B** The number in Column II is greater.
**C** The two numbers are equal.
**D** The relationship cannot be determined from the given information.

| Column I | Column II |
|---|---|
| 6. the y-coordinate of any point on the graph of $y = -\sqrt{x}$ except 0 | the y-coordinate of any point on the graph of $y = \sqrt{x}$ except 0 |
| Ⓐ    Ⓑ | Ⓒ    Ⓓ |
| 7. the value of x when the point $(-3, 2)$ is reflected across the y-axis | the value of x when the point $(-3, 2)$ is reflected across the x-axis |
| Ⓐ    Ⓑ | Ⓒ    Ⓓ |
| 8. the x-coordinate of the vertex of $y = x^2$ | the x-coordinate of the vertex of $y = -x^2$ |
| Ⓐ    Ⓑ | Ⓒ    Ⓓ |

9. **Multiple Choice** The graph below is the reflection of the graph of $y = -3x + 1$ across the:

   Ⓐ y-axis.
   Ⓑ point $(-1, 1)$.
   Ⓒ origin.
   Ⓓ x-axis.

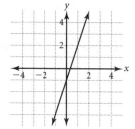

Algebra 1                                              Standardized Test Practice 14.4    **99**

NAME _____ CLASS _____ DATE _____

# Standardized Test Practice
## 14.5 Combining Transformations

**TEST TAKING STRATEGY** Treat multiple choice questions like they are true/false questions.

1. **Multiple Choice** The graph of $y = x^2$ is translated 2 units to the right and reflected across the *x*-axis. Which is an equation for the transformation?

   Ⓐ $y = -x^2 - 2$
   Ⓑ $y = -(x + 2)^2$
   Ⓒ $y = -(x - 2)^2$
   Ⓓ $y = -x^2 + 2$

2. **Multiple Choice** Which describes the transformation of the graph of $y = |x|$ to the graph of $y = |0.05x| - 2$?

   Ⓐ a horizontal stretch by a factor of 2 followed by a vertical translation 2 units down
   Ⓑ a horizontal compression by a factor of 0.5 followed by a vertical translation 2 units down
   Ⓒ a vertical stretch by a factor of 2 followed by a vertical translation 2 units down
   Ⓓ a vertical compression by a factor of 0.5 followed by a vertical translation 2 units down

3. **Multiple Choice** A transformation of the graph of $y = \sqrt{x}$ combines a reflection across the *y*-axis, a vertical stretch by a factor of 2, and a reflection across the *x*-axis. Which is an equation for the transformation?

   Ⓐ $y = -\sqrt{-2x}$
   Ⓑ $y = -2\sqrt{-x}$
   Ⓒ $y = 2\sqrt{-x}$
   Ⓓ $y = -2\sqrt{x}$

**Quantitative Comparison** In Exercises 4–5, choose the letter of the statement below that is true about the quantities in Columns I and II.

A The number in Column I is greater.
B The number in Column II is greater.
C The two numbers are equal.
D The relationship cannot be determined from the given information.

| | Column I | Column II |
|---|---|---|
| 4. | the value of $a$ in $y = af(x)$ if the graph represents a vertical stretch | the value of $h$ in $y = f(x - h)$ if the graph represents a horizontal translation to the left |
| | Ⓐ  Ⓑ | Ⓒ  Ⓓ |
| 5. | the value of $y$ in $(1, y)$ if the graph of $y = x^2$ is translated right 1 unit | the value of $y$ in $(1, y)$ if the graph of $y = 2x^2$ is reflected across the *y*-axis |
| | Ⓐ  Ⓑ | Ⓒ  Ⓓ |

6. **Multiple Choice** The graph shown is a transformation of the graph of $y = |x|$. Which equation represents the transformation?

   Ⓐ $y = |-2x + 3|$
   Ⓑ $y = -|-2x - 3|$
   Ⓒ $y = |2x - 3|$
   Ⓓ $y = -|2x + 3|$

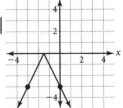

NAME _____ CLASS _____ DATE _____

# SAT/ACT Chapter Test

## Chapter 14  Functions and Transformations

**TEST TAKING STRATEGY**  Be sure you understand what is being asked in the question.

1. **Multiple Choice** The graph of $y = |x|$ is translated right 2 units and up 3 units. The equation of the translated graph is:

   Ⓐ $y = |x + 2| + 3$
   Ⓑ $y = |x + 5|$
   Ⓒ $y = |x - 2| + 3$
   Ⓓ $y = |x + 2| - 3$

2. **Multiple Choice** The graph of $y = -x^2 - x - 2$ is reflected across the x-axis. Which is an equation for the graph?

   Ⓐ $y = x^2 + x - 2$
   Ⓑ $y = x^2 + x + 2$
   Ⓒ $y = -x^2 + x - 2$
   Ⓓ $y = (-x)^2 + x - 2$

3. **Multiple Choice** Which describes the transformation of the graph of $y = x^2$ to the graph of $y = -3(x + 2)^2 + 1$?

   Ⓐ a horizontal translation 2 units right, followed by a vertical stretch by a factor of 3, followed by a reflection across the x-axis, followed by a vertical translation up 1 unit

   Ⓑ a horizontal translation 2 units left, followed by a vertical stretch by a factor of 3, followed by a reflection across the x-axis, followed by a vertical translation up 1 unit

   Ⓒ a reflection across the x-axis, followed by a translation up 1 unit

   Ⓓ a vertical stretch by a factor of 3, followed by a reflection across the x-axis

**Quantitative Comparison** In Exercises 4–6, choose the letter of the statement below that is true about the quantities in Columns I and II.

**A** The number in Column I is greater.
**B** The number in Column II is greater.
**C** The two numbers are equal.
**D** The relationship cannot be determined from the given information.

| | Column I | Column II |
|---|---|---|
| 4. | the value of $h$ if $y = f(x - h)$ represents a shift of 3 units to the right | the value of $b$ if $y = f(bx)$ represents a horizontal stretch by a factor of 3 |
| | Ⓐ  Ⓑ  Ⓒ  Ⓓ | |
| 5. | the value of $k$ if $y = f(x)$ is translated vertically $k$ units | the value of $h$ if $y = f(x)$ is translated horizontally $|h|$ units |
| | Ⓐ  Ⓑ  Ⓒ  Ⓓ | |
| 6. | the value of $y$ in $(2, y)$ if the point is on the graph of $y = |-3x| - 6$ | the y-coordinate of the vertex of the parabola $y = -100x^2$ |
| | Ⓐ  Ⓑ  Ⓒ  Ⓓ | |

7. **Multiple Choice** If $f(x) = -x^2 + 3x - 5$, what is $f(-2)$?

   Ⓐ 7
   Ⓑ 3
   Ⓒ 15
   Ⓓ −15